Pharmacy Technician Certification Exam Review

version 1.0

by
Sean E. Parsons, CPhT

P³

Parsons Printing Press
328 Janice Drive
Pittsburgh, PA 15235
sean@parsonsprintingpress.com

Copyright

Acknowledgments

As with any book, nothing exists in a vacuum; therefore, many people deserve acknowledgment for this book, probably more than I will ever be able to recognize. I want to give thanks to the Pharmacy Technician Certification Board (PTCB) for the many years that they have spent elevating the profession to what it is today, even though there are many things that are still needed to advance this profession further. I want to thank all the pharmacy technician instructors that have dedicated much of their careers training the highly educated pharmacy technicians that we have today and the technicians that will come tomorrow. I want to thank the school where I teach, Bidwell Training Center, Inc., for giving me the opportunity to share my passion for pharmacy with others.

A more personal thanks belongs to my editor, Marsha Posey who had the laborious task of correcting both my grammar and my spelling. If any sentence in this book is strung together in an appropriate manner, it is due to her diligence.

I want to thank Kelly Gambridge and Anna Ross, as they also had a very cumbersome task. Kelly and Anna entered all the information from this book into a database to help create the etools that accompany this book.

Last, but certainly not least, I want to thank Shannon, my better three-quarters (calling her my better half would seriously undervalue her). She is the most supportive spouse one could ever imagine, as she allows me to do every peculiar thing my heart desires (writing books, building websites, developing software, starting a publishing company, etc.).

Sean Parsons, CPhT

How to Use This Book

This book is broken up topically, and as this book is intended as a study aid for the *Pharmacy Technician Certification Exam* (PTCE), you may choose which order you want to review the material. The introduction also includes a lot of information about the exam itself, such as how much each subject on the exam is emphasized, what you need to do to register for the exam, and even instructions to help you find out where you can take the exam.

Whether you are looking at an e-version of this book, or the more familiar dead-tree format, you will find questions posed throughout each study chapter attempting to reinforce the material that was just reviewed. At the end of each chapter is a list of the correct responses along with explanations in order to help you better understand the correct responses.

There are also free online practice questions available at http://www.parsonsprintingpress.com/etools/. The site does require free registration, and is available to anyone even if they have not purchased this book. The site has been optimized to work well with both traditional desktop computers and mobile devices. This may be useful if you are working on a practice quiz at home on your laptop and later want to finish a particular practice exam, you could log back in on your smart phone.

After you have successfully passed the PTCE, this book will continue to be useful as it will provide information about acquiring continuing education and how to renew your certification.

Study hard and good luck on passing the certification exam.

Table of Contents

Introduction

Some pharmacy technicians have only on-the-job training, but many employers favor those who have completed a formal training and/or certification process. In order to make yourself more marketable as a pharmacy technician, it is strongly recommended that you become certified. The largest national certification exams in the United States are given by the Institute for the Certification of Pharmacy Technicians (ICPT) and the Pharmacy Technician Certification Board (PTCB). PTCB is the pioneer and industry leader in pharmacy technician certification. Founded in 1995, PTCB has certified over 450,000 pharmacy technicians. The exam given by the PTCB is the Pharmacy Technician Certification Examination (PTCE) and is the only pharmacy technician examination endorsed by the American Pharmacists Association (APhA), the American Society for Health-System Pharmacists (ASHP), and the National Association of Boards of Pharmacy (NABP). The exam given by ICPT is called the Exam for the Certification of Pharmacy Technicians (ExCPT). Completing one of these exams earns the technician the credentials "CPhT", corresponding to the professional title of Certified Pharmacy Technician.

Since the PTCE is the more recognized certification, we will concentrate on the appropriate preparations for this exam, but if you are so inclined to pursue the ExCPT most of the material in this book should also be valid to prepare you for that exam as well.

Book overview

This book is broken into nine subject areas proceeded by practice exams and steps to follow after certification. The nine subject areas include:

- pharmacology for technicians,
- pharmacy law and regulation,
- sterile and non-sterile compounding,
- medication safety,
- pharmacy quality assurance,
- medication order entry and fill process,

- pharmacy inventory management,
- pharmacy billing and reimbursement, and
- pharmacy information system usage and application.

Various questions will be asked throughout the chapters, as well as the practice exam at the end of this book, to help readers prepare for the certification exam. Answers to all questions in every chapter will be included at the end of the chapter.

Frequently asked questions

These questions and responses will be based on the Pharmacy Technician Certification Examination (PTCE) as it is the more recognized of the certification exams mentioned above.

What is the PTCB?

The PTCB is a national certification program that enables pharmacy technicians to work more effectively with pharmacists to offer safe and effective patient care and service. PTCB develops and implements policies related to national certification for pharmacy technicians. PTCB administers a nationally accredited certification examination for pharmacy technicians which tests knowledge in all practice settings.

What is certification?

Certification is the process by which a non-governmental association or agency grants recognition to an individual who has met certain predetermined qualifications specified by that association or agency.

What is the recognition of certification?

Individuals who meet all eligibility requirements and who successfully pass the national Pharmacy Technician Certification Examination may use the designation "CPhT" after their name. A certificate and wallet card will be sent to newly Certified Pharmacy Technicians approximately six to eight weeks after sitting for the examination. Certification is valid for two years.

If I am certified as a pharmacy technician in one state will that cover me for all states?

PTCB's certification is a national certification that is valid nationwide. However, the regulations to work in a pharmacy as a pharmacy technician vary from state to state. Contact your state board of pharmacy for whatever state you would prefer to practice in or visit the National Association of Boards of Pharmacy at http://www.nabp.net.

When is the test offered?

As of April 1, 2009, the PTCB has been offering continuous testing.

Where will the exam be administered?

PTCB will offer the exam through Pearson Vue's extensive network of over 200 test center locations in all 50 of the states. The web link to check testing locations is http://pearsonvue.com/vtclocator/.

How much does the exam cost?

Effective February 5th, 2007, the fee for the PTCE exam is $129.00

How do I register for the exam?

Visit http://www.ptcb.org to register for the Pharmacy Technician Certification Examination (PTCE) online or by calling 866-902-0593.

Once an exam appointment is scheduled, can it be changed or even canceled?

Yes. Candidates can cancel or reschedule their exam appointment at a minimum of 24 hours prior to their scheduled time at no charge.

What is on the exam?

As of November 1, 2013, the content of the exam is characterized under nine function areas:

- Pharmacology for Technicians - 13.75%
- Pharmacy Law and Regulations - 12.5%
- Sterile and Non-Sterile Compounding - 8.75%
- Medication Safety - 12.5%
- Pharmacy Quality Assurance - 7.5%
- Medication Order Entry and Fill Process - 17.5%
- Pharmacy Inventory Management - 8.75%
- Pharmacy Billing and Reimbursement - 8.75%

- Pharmacy Information System Usage and Application - 10%

Those concepts, while very broad in total scope, are each further defined by the PTCB and are very understandable for a candidate if they have either formal education or adequate pharmacy experience. Each of these function areas will be reviewed throughout this book.

How many questions are on the exam?

The PTCE contains 90 multiple-choice questions, but only 80 of which are scored. The other 10 questions, scattered throughout the exam, are used for building future exams. Candidates are encouraged to answer all questions. Each question provides four choices, with only ONE designated as the correct or best answer. The questions from the nine functions tested are distributed randomly throughout the total exam. It is to your advantage to answer every question on the exam, since the final score is based on the total number (of the 80 scored questions) answered correctly.

How long is the exam?

Candidates will have 1 hour 50 minutes to complete the exam. Prior to the actual exam will be a short on-screen tutorial, and after the exam there will be a brief exit survey.

Can I use a calculator?

Candidates are permitted to use the on-screen calculator, or a handheld calculator offered by the testing site.

What should I do on the day of the exam?

- Arrive at the test center 30 minutes prior to the start of your examination.
- Bring a clear, legible, and valid government-issued photo identification.
- Bring your Authorization to Test letter with you to the testing center. (Your name on the ID and Authorization to Test letter must match.)

What happens on the day of the examination?

PTCB recommends that candidates arrive at their chosen Pearson Professional Center 30 minutes prior to the scheduled start time of their examination. Upon arrival, the candidate will sign in with the test center staff and store their personal belongings in lockers provided by Pearson VUE (no personal belongings are permitted in the testing room). Candidates will be provided with a one-page eraser board or plastic sheet to serve as scratch paper. If you need additional scratch paper, you may raise your hand and an exam proctor will collect your full board/sheet and then provide you with an additional marker board or plastic sheet. Also, you will be given the option of using ear plugs provided by the examination center while you test.

What does the exam actually look like?

When you are taken to the computer that your exam will be conducted on, a Pearson Vue associate will sign you into your test. The computer will offer a brief tutorial on how to use the test and then the test will begin. The computer screen will look similar to the image below:

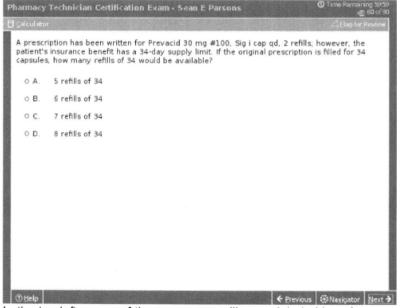

In the top left corner of the exam, you will see a labeled icon for calling up an on screen calculator (if it is a math question). In the top right corner of the exam, you can see how much time you have remaining, which question you're on out of your total number of questions, and the ability to flag a question in order to review it later. In the bottom left, you will find a *Help* icon that will provide reminders on how to navigate the exam. On the bottom right of the exam, you will find the ability to navigate to the previous or the next question along with a navigator control that will let you jump to any question, including showing which ones you haven't answered and which questions you flagged. At the end of the exam, provided you have time left, you will also be given an opportunity to review any question, including showing which ones you haven't answered and which questions you flagged.

When will I know if I passed?

Unofficial scores will be made available at the end of the exam. Official score reports will be mailed and available online within two weeks of a candidate completing the PTCE.

When can I use the designation CPhT?

Once you successfully pass the PTCE and receive your official scores, you may use the designation of CPhT (Certified Pharmacy Technician) following your name. A certificate and wallet card will be sent to newly Certified Pharmacy Technicians approximately 60 days after sitting for the certification examination.

Will I have to complete continuing education courses?

Once certification is achieved, recertification is required every two years. A total of 20 hours of Continuing Education Units (CEUs or CE) of technician-appropriate continuing education in pharmacy-related topics is required within the two-year period. At least one of the 20 hours of CE must be in the area of pharmacy law during each renewal period, and starting in 2014 one of the CE must be in medication safety during each renewal period. In 2015, all CE will need to be pharmacy technician specific (until then, CE designated for pharmacists can still be used by technicians).

Acceptable pharmacy technician continuing education can be obtained through a variety of methods, such as through pharmacy or pharmacy technician professional organizations, as well as through a number of sources online.

You may use 1 college course during your two-year certification period. The college course must be in either a Life Science (Chemistry, Biology, Anatomy, etc.) or Math, and currently counts as 15 hours of CE. That value will be reduced to 10 hours of CE in 2016.

More about certification renewal will be covered in the last chapter of this book.

CHAPTER 1
Pharmacology for Technicians

Key concepts

This chapter will cover the following knowledge areas to prepare you for the *Pharmacy Technician Certification Exam*:

- Generic and brand names of pharmaceuticals
- Therapeutic equivalence
- Drug interactions (e.g., drug-disease, drug-drug, drug-dietary supplement, drug-OTC, drug-laboratory, drug-nutrient)
- Strengths/dose, dosage forms, physical appearance, routes of administration, and duration of drug therapy
- Common and severe side or adverse effects, allergies, and therapeutic contraindications associated with medications
- Dosage and indication of legend, OTC medications, herbal and dietary supplements

Terminology

To get started in this chapter, there are some terms that should be defined.

pharmacology - The term pharmacology is derived from two Greek words "pharmakon" and "logos". Pharmakon can mean sacrament, remedy, poison, talisman, cosmetic, perfume or intoxicant, but in this case, it can be broadly defined as drug. Logos can be translated as a principle of order and knowledge. By combining the terms you can see that pharmacology is concerned with the knowledge of drugs.

OTC - Over the counter medications are drugs that do not require the use of a prescription for a patient to obtain it. These medications are generally considered sufficiently safe for a patient to acquire and self medicate with by following the instructions included on the vial.

legend - Legend drugs are medications that require a prescription prior to dispensing. These medications may or may not be considered abusable, but consultation with a medical professional that has prescribing authority is necessary due to the complex health conditions that these medications may be able to treat or ameliorate. Often, you will see the phrase "Rx only" used to denote legend drugs.

controlled substances - Controlled substances are medications with restrictions due to abuse potential. There are 5 schedules of controlled substances with various prescribing guidelines based on abuse potential counter balanced by potential medicinal benefit as determined by the Drug Enforcement Administration and individual state legislative branches.

brand name - This is the manufacturer's trademark designation. Each brand name is owned by the company and begins with a capital letter, and it is protected by a trademark. Drugs often have several brand names. The terms trade name and proprietary name may also be used interchangeably with brand name.

generic name - This is the official non-proprietary name assigned by the manufacturer with the approval of the USAN (United States Adopted Name) Council. The Federal Food and Drug Administration (FDA) requires that each drug has a generic name, even if currently it is only available as a brand name product. A generic name may also be referred to as the non-proprietary name.

therapeutic equivalents - Drug products are considered to be therapeutic equivalents only if they are available in the same dosage strength, dosage form, achieve the same blood levels, and if they can be expected to have the same clinical effect and safety profile when compared to the original innovator drug. To ensure that a specific generic product is considered to be a therapeutic equivalent, it is recommended that you refer to the Orange Book, a publication from the Food and Drug Administration (FDA). If the products being compared in the Orange Book are considered therapeutic equivalents, it will be given a Therapeutic Equivalence (TE) Code of "A". If the products are not considered therapeutic equivalents, they will be given a TE Code of "B". Sometimes these will be referred to as "A" ratings and "B" ratings.

classification - This is how a medication is grouped and is typically defined according to its use in treating a particular disease or disorder. A potential source of confusion is that many medications could be grouped multiple ways. An example of this is aspirin which can be classified as an analgesic, antipyretic, and anti-inflammatory agent.

indication - This is the primary condition(s) treated by a particular drug. This may include both FDA approved uses as well as off-label (generally based on scientific studies but lacking FDA approval) medication uses.

dosage form - A dosage form is the physical form of a dose of medication, such as capsule, patch, or injection. The route of administration is dependent on the dosage form of a given drug. As an example, persistent vomiting may make it difficult to use an oral dosage form; therefore, an antiemetic in suppository form may be desirable.

interactions - Medications have the potential to interact with other medications, dietary supplements, and constituents of various foods. Medications may also interact with various diseases.

Quick review question 1

Which of the following is not a synonym for the term brand name?

  a. trade name
  b. proprietary name
  c. non-proprietary name

Quick review question 2

What does it mean if you compare two products in the Orange Book and get a TE Code of "B"?

  a. The two medications are considered therapeutic equivalents and can be safely interchanged with each other.
  b. There is a therapeutic equivalence problem.
  c. The generic product has a grade range of 80-89%.

d. There is no such thing as a TE Code of "B".

Common medications

Pharmacy technicians should be able to categorize drugs into major therapeutic classifications and be able to briefly describe the therapeutic use of each drug. Technicians should also be able to correlate brand name medications with the generic names.

The following subsections will break the drugs down into various therapeutic classes, with descriptions of these classes followed by a list of generic drug names accompanied by their most common brand names and commercially available dosage forms. These lists are not to be considered comprehensive, but instead just provide some of the more common items to serve as a quick refresher. If you want more information on a particular drug, an excellent free (registration required) resource is Medscape at http://reference.medscape.com.

Anti-infective agents

An anti-infective (also called an antimicrobial) is a substance that kills or inhibits the growth of microbes such as bacteria, fungi, protozoans or viruses. In this section we will look at antibiotics, antifungals, amebicides, and antivirals.

Anti-infectives are usually given intermittently for a short period of time (3 to 14 days), although there are exceptions to this.

Antibiotics

Bacteria treated by antibiotics are usually broken into two categories, gram-positive and gram-negative.

Gram-positive bacteria are those that are stained dark blue or violet by Gram staining. This is in contrast to gram-negative bacteria, which cannot retain the crystal violet stain, instead taking up the counterstain (safranin or fuchsin) and appearing red or pink. Gram-positive organisms are able to retain the crystal violet stain because of the high amount of peptidoglycan in the cell wall. Gram-positive cell walls typically lack the outer membrane found in gram-negative bacteria.

Penicillin Derivatives

Penicillin derivatives act by preventing cell wall synthesis during active replication and are therefore bactericidal. Penicillin derivatives have a chemical structure referred to as a beta-lactam ring, which is what allows it to attach to the bacteria. Penicillins are used in the treatment of bacterial infections caused by susceptible, usually gram-positive, organisms. The most common side effects are allergic reactions including rash, hives, or anaphylactic shock.

amoxicillin (Amoxil, Moxatag, Trimox) - oral solution, capsule, tablet, chewable tablet, extended-release tablet
ampicillin (Principen) - capsules, oral suspension, powder for injection
nafcillin (Nafcil) - injectable solution, powder for injection
oxacillin (Bactocill) - infusion solution, oral solution, powder for injection
penicillin G (Bicillin, Wycillin) - premixed injectable solution, powder for injectable solution
penicillin V potassium (Pen Vee K, Veetids) - oral solution, tablet
piperacillin (Pipracil) - powder for injection

Penicillin Derivatives with Beta-Lactamase Inhibitors

Sometimes bacteria become resistant to penicillin. One such mechanism of action for them to do that is by creating beta-lactamase, which breaks down the beta-lactam ring in penicillins preventing them from attaching to the bacteria walls. There is a chemical class that can prevent this called beta-lactamase inhibitors that are often coupled with penicillins.

amoxicillin & clavulanate (Augmentin) - oral suspension, tablet, chewable tablet
ampicillin & sulbactam (Unasyn) - injectable solution, powder for solution
piperacillin & tazobactam (Zosyn) - powder for injection
ticarcillin & clavulanate (Timentin) - powder for injection, infusion solution

Carbapenems

Carbapenems are a class of beta-lactam antibiotics with a broad spectrum of antibacterial activity, and have a structure which renders them highly resistant to beta-lactamases.

ertapenem (Invanz) - powder for injection
imipenem & cilastatin (Primaxin) - powder for injection
meropenem (Merrem) - powder for injectable solution

Cephalosporins

The cephalosporins are bactericidal antibiotics that have chemical structures similar to those of penicillins and therefore have the same mechanism of action. Also, due to their similar chemical structure to penicillin, there is a 7-10% potential of cross sensitivity. The cephalosporins are considered to be broad spectrum drugs. Their two main uses are as substitutes for penicillins in cases of allergy or bacterial resistance and in the treatment of certain gram-negative infections.

cefaclor (Ceclor) - capsule, tablet extended-release
cefazolin (Kefzol) - powder for injection
cefdinir (Omnicef) - capsule, oral suspension
cefepime (Maxipime) - infusion solution, powder for injection
cefotaxime (Claforan) - injectable solution, powder for injection
cefotetan (Cefotan) - injectable solution, powder for injection
ceftaroline (Teflaro) - powder for injection
ceftazidime (Fortaz) - injectable solution, powder for injection
ceftriaxone (Rocephin) - injectable solution, powder for injection
cephalexin (Keflex) - capsule, film-coated tablet, oral suspension, tablet for oral suspension

Aminoglycosides

The aminoglycosides are a group of bactericidal antibiotics whose antibacterial spectrum mainly includes gram-negative bacilli. Aminoglycosides have poor oral absorption; therefore, if systemic effects are desired, they are usually given IV or IM. Aminoglycosides are both nephrotoxic (kidney) and ototoxic (ear).

amikacin (Amikin) - injectable solution
gentamicin (Garamycin) - injectable solution, ophthalmic ointment, ophthalmic suspension
kanamycin (Kantrex) - injectable solution
tobramycin (Nebcin, Tobi, Tobrex) - injectable solution, nebulizer

solution, ophthalmic ointment, ophthalmic solution
tobramycin & dexamethasone (TobraDex) - ophthalmic ointment, ophthalmic suspension - This is a combination of an aminoglycoside and a corticosteroid.

Tetracyclines

The tetracyclines (a group of broad-spectrum bacteriostatic antibiotics) are clinically useful in both gram-positive and gram-negative infections.

When ingested, it is usually recommended that tetracyclines should be taken with a full glass of water, either two hours after eating or one hour before eating. This is partly because tetracycline binds easily with magnesium, aluminum, iron, and calcium, which reduces its ability to be completely absorbed by the body. Dairy products or preparations containing iron are not recommended directly after taking the drug.

Side effects from tetracyclines are not always common; but of particular note is possible photosensitive allergic reaction which increases the risk of sunburn under exposure to UV light from the sun or other sources. Tetracyclines are teratogens due to the likelihood of causing teeth discoloration in the fetus as they develop in infancy. For this same reason, tetracyclines are contraindicated for use in children under 12 years of age.

doxycycline (Vibramycin) - capsule, powder for injection, syrup, tablet, delayed-release tablet
minocycline (Minocin) - oral suspension, tablet, capsule, extended-release tablet, intravenous injection
tetracycline (Sumycin) - capsule, tablet
tigecycline (Tygacil) - powder for injection

Sulfonamides

Sulfonamides are a group of synthetic bacteriostatic drugs which are effective against both gram-positive and gram-negative infections. Approximately 3% of people using sulfonamides report side effects, the most common of which are hypersensitivity reactions such as rash and hives, but some patients may have more severe reactions.

Sulfonamides have the potential to crystallize in the kidneys, due to their low solubility. This is a very painful experience, so patients are recommended to take these medications with large amounts of water.

silver sulfidiazine (Silvadene) - cream
sulfamethoxazole & trimethoprim, SMZ &TMP (Bactrim, Septra, Co-Trimoxazole) - injected solution, oral solution, tablet
sulfasalazine (Azulfidine) - tablet, extended-release tablet

Macrolide Antibiotics

The term macrolide refers to the large chemical ring structure that is characteristic of these antibiotics. These antibiotics inhibit bacterial protein synthesis and can be bacteriostatic or bactericidal. Food affects oral absorption of these drugs, although they frequently cause GI problems which may warrant taking them with food. Macrolides may interfere with some medications that require liver biotransformation such as digoxin, warfarin, and cyclosporine; therefore, a different antibiotic choice may be desirable if a patient is currently using any of those medications.

azithromycin (Zithromax, AzaSite) - oral suspension, powder for injection, powder for oral suspension, extended-release powder for oral suspension, tablet, ophthalmic solution
clarithromycin (Biaxin) - oral suspension, tablet, extended-release tablet
erythromycin (Erythrocin, E-Mycin) - tablet, delayed-release tablet, dispertab, injection, oral suspension, ophthalmic ointment

Fluoroquinolones

The fluoroquinolones, often just called quinolone antibiotics, are synthetic antimicrobial agents that are a broad spectrum bactericidal, especially against gram-negative organisms. A black box warning has been added to all fluoroquinolones involving an increased risk of developing tendinitis and tendon rupture in patients of all ages taking fluoroquinolones for systemic use. This risk is further increased in individuals over 60 years of age, taking corticosteroid drugs, and having received kidney, heart, or lung transplants.

ciprofloxacin (Cipro, Ciloxan) - infusion solution, oral suspension, tablet, extended-release tablet, ophthalmic solution, ophthalmic ointment, otic solution

levofloxacin (Levaquin) - injectable solution, premix infusion solution, oral solution, tablet, ophthalmic solution
moxifloxacin (Avelox) - injectable solution, tablet, opthalmic solution
ofloxacin (Floxin) - tablet, ophthalmic solution, otic solution

Miscellaneous Antibiotics

Some common antibiotics are the only drugs in their respective classes. The following listed medications are examples of those. These drugs each have their own side effects and precautions; for example, IV use of vancomycin may cause a reaction known as red man syndrome characterized by flushing and/or rash that affects the face, neck, and upper torso.

aztreonam (Azactam) - infusion solution, powder for injection
clindamycin (Cleocin) - capsule, injectable solution, oral solution, vaginal cream, vaginal suppository
daptomycin (Cubicin) - injectable solution
linezolid (Zyvox) - injectable solution, oral suspension, tablet
mupirocin (Bactroban) - cream, topical ointment, nasal ointment
nitrofurantoin (Macrobid) - capsule, oral suspension
vancomycin (Vancocin) - capsule, injectable solution, powder for injection

Antifungals

An antifungal drug is a medication used to treat fungal infections such as athlete's foot, ringworm, candidiasis (thrush), serious systemic infections such as cryptococcal meningitis, and others.

Antifungals work by exploiting differences between mammalian and fungal cells to kill off the fungal organism without dangerous effects on the host. Unlike bacteria, both fungi and humans are eukaryotes. Thus fungal and human cells are similar at the molecular level. This means it is more difficult to find a target for an antifungal drug to attack that does not also exist in the infected organism.
Consequently, there are often side-effects to some of these drugs. Many of these agents are hepatotoxic (liver), and liver function and enzymes must be monitored. It is not uncommon for therapy to last for several months.

amphotericin B (Fungizone) - powder for injection, injectable lipid complex, cream, lotion, ointment

betamethasone & clotrimazole (Lotrisone) - cream, lotion - This is a combination of a corticosteroid and an antifungal.
clotrimazole (Gyne-Lotrimin, Mycelex) - troche, vaginal cream, vaginal tablet, cream, topical solution, lotion, powder
fluconazole (Diflucan) - injected solution, oral suspension, tablet
nystatin (Mycostatin, Nilstat) - oral powder, oral suspension, oral tablet, troche, cream, ointment, powder, vaginal tablet
terbinafine (Lamisil) - tablet, oral granules, cream, topical solution

Amebicides

An amebicide (or amoebicide) is an agent used in the treatment of amoebic infections. Alcohol use should be avoided while taking metronidazole because concurrent use may cause nausea, vomiting, flushing of the skin, tachycardia, and shortness of breath (this is referred to as a disulfarim-like reaction).

metronidazole (Flagyl, MetroGel) - capsule, tablet, extended-release tablet, infusion solution, topical gel, topical cream, topical lotion, vaginal gel
pentamidine (Pentam) - powder for injection, powder for nebulizer solution

Antimalarial Drugs

Antimalarial drugs are agents used to prevent and cure malaria. Some antimalarial drugs, such as hydroxychloroquine, are also used to treat conditions like rheumatoid arthritis and lupus. One of the most serious side effects is a toxicity in the eye (this primarily occurs with chronic use).

hydroxychloroquine (Plaquenil) - tablet
quinine sulfate (Qualaquin) - capsule

Antivirals

Antiviral drugs are a class of medication used specifically for treating viral infections. Like antibiotics, specific antivirals are used for specific viruses. They are relatively harmless to the host, and therefore can be used to treat infections. They should be distinguished from viricides, which actively deactivate virus particles outside the body.

Most of the antivirals now available are designed to help deal with

HIV (ganciclovir); herpes viruses (acyclovir, ganciclovir, valacyclovir), best known for causing cold sores and genital herpes, but actually causing a wide range of diseases; the hepatitis B and C viruses, which can cause liver cancer; and influenza viruses (oseltamivir). Researchers are now working to extend the range of antivirals to other families of pathogens.

Antiviral drugs work by inhibiting the virus either before it enters the cell, stopping it from reproducing, or in some cases, preventing it from exiting the cell. However, like antibiotics, viruses may evolve to resist the antiviral drug.

acyclovir (Zovirax) - injectable solution, oral suspension, powder for injection, tablet, capsule, topical cream, topical ointment
ganciclovir (Cytovene) - capsule, powder for injection, oral solution, implant, ophthalmic gel
oseltamivir (Tamiflu) - capsule, powder for oral suspension
valacyclovir (Valtrex) - tablet

Quick review question 3

A patient receiving 100 mg of doxycycline every 12 hours should avoid taking their medication with which of the following items?

   a. milk
   b. food
   c. Maalox (magnesium aluminum hydroxide)
   d. all of the above

Quick review question 4

If a patient has a penicillin allergy, which of the following antibiotics should they avoid?

   a. Keflex
   b. Vancocin
   c. Septra
   d. Minocin

Quick review question 5

What is the generic name for Zosyn?

   a. amoxicillin & clavulanate
   b. ampicillin & sulbactam

c. piperacillin & tazobactam
d. ticarcillin & clavulanate

Quick review question 6

Zovirax (acyclovir) would be most accurately classified as which of the following?

a. antibiotic
b. amebicide
c. antiviral
d. viricide

Quick review question 7

A prescription for sulfamethoxazole & trimethoprim should receive which of the following auxiliary labels?

a. Take medication on an EMPTY STOMACH
b. Medication should be taken with plenty of WATER
c. Do not take with dairy products, antacids, or iron preparations
d. Take with FOOD

Glucocorticosteroids

Glucocorticosteriods (glucocorticoids) are steroids used to treat inflammation and various allergic conditions. They are similar to the steroid created in the adrenal cortex, although they could be either from natural sources or synthetically manufactured. As your body naturally produces a glucocorticoid, introducing an exogenous source through medication interferes with your body's negative feedback loop for production of these steroids. This is why it is often necessary to taper a patient off of a steroid so their body slowly starts creating its own endogenous source of steroids again.

Long term use of glucocorticoids can have the following negative effects: thinning of skin, decrease wound healing, stunting pediatric growth, moon-face, obesity, and diabetes mellitus.

Short-Acting Steroids

hydrocortisone (Cortef) - tablet, oral suspension, powder for injection, rectal cream, enema, foam, suppositories, topical cream, lotion, gel, topical solution, ointment, pledget

Intermediate-Acting Steroids

fluticasone (Flovent HFA, Flonase) - aerosol inhaler, disk inhaler, nasal spray, cream, ointment, lotion
methylprednisolone (Depo-Medrol, Medrol, Solu-Medrol) - tablet, injectable suspension, powder for injection
prednisolone (Delta-Cortef) - oral solution, tablet, ophthalmic suspension, ophthalmic solution
prednisone (Deltasone) - oral solution, tablet
triamcinolone, TAC (Azmacort, Nasocort AQ, Kenalog) - nasal spray, inhalation aerosol, intravitreal injection, cream, ointment, paste, topical spray

Long-Acting Steroids

dexamethasone (Decadron) - tablet, injectable suspension, elixir, oral solution, oral concentrate, ointment, ophthalmic solution, ophthalmic suspension, intravitreal implant

Quick review question 8

Which of the following is a potential side effect of long-term use of glucocorticoids?

   a. thinning of skin
   b. decreased wound healing
   c. moon-face
   d. all of the above

Drugs affecting the sympathetic nervous system

The sympathetic nervous system is part of the autonomic nervous system that tends to act in opposition to the parasympathetic nervous system, by speeding up the heartbeat and causing contraction of the blood vessels. It regulates the function of the sweat glands and stimulates the secretion of glucose in the liver. The sympathetic nervous system is usually activated under conditions of stress, which causes the 'fight or flight' response.

The sympathetic nervous system can be excited or inhibited through either stimulating or inhibiting adrenergic receptors. These adrenergic receptors are normally stimulated by endogenous norepinephrine, but they can be stimulated by exogenous chemicals of either natural or synthetic origin. These adrenergic receptors can be further broken up into subgroups called alpha-1, alpha-2, beta-1, and beta-2 adrenergic receptors. Excitation of alpha-1 and alpha-2 adrenergic receptors causes contraction of smooth muscle resulting in vasoconstriction of most blood vessels, contraction of sphincter muscles in the GI and urinary tract, and dilation of the pupil of the eye (mydriasis). Stimulation of beta-1 adrenergic receptors cause stimulation of the heart (increased heart rate and force of contraction). Stimulation of beta-2 adrenergic receptors cause bronchodilation. Inhibition (blocking) of these receptors cause the opposite effect to occur.

Let's look at drugs that affect these various receptors.

Adrenergic Agonists

These drugs stimulate all the adrenergic receptors to varying degrees and will therefore have the effects related to stimulation of these receptors. This allows these drugs to be used for a multitude of purposes, such as acute hypotension (ephedrine, NE, PE), cardiac arrest (EPI, NE), severe asthma (EPI), bronchodilation (ephedrine), anaphylaxis (EPI), congestion (PE), and can be used to dilate pupils (PE).

ephedrine - injectable solution
epinephrine, EPI (EpiPen, Racepinephrine) - autoinjector, injectable solution, solution for nebulization
norepinephrine, NE (Levophed) - injectable solution
pseudoephedrine, PE (Sudafed) - tablet, syrup

Alpha-2 Adrenergic Agonists

Alpha-2 agonists, despite stimulating the alpha-2 adrenergic receptors, actually function against the sympathetic nervous system and sometimes will be classified as sympathetic blocking (sympatholytic) drugs. The net result is a decrease in cardiac output and vasodilation making them useful in the treatment of hypertension.

clonidine (Catapres, Catapres TTS) - injectable solution, extended-release oral suspension, patch, tablet, extended-release tablet
guanfacine (Tenex) - tablet

Alpha-Adrenergic Blocking Agents

Alpha-adrenergic blocking drugs are primarily used to treat benign prostatic hypertrophy (BPH). These medications also cause vasodilation and reduced blood pressure, making them suitable choices for treating hypertension.

doxazosin (Cardura, Cardura XL) - tablet, extended-release tablet
tamsulosin (Flomax) - capsule
terazosin (Hytrin) - capsule

5-Alpha-Reductase Inhibitors

5-alpha-reductase inhibitors are a group of medications with antiandrogenic activity, and are used in benign prostatic hypertrophy (dutasteride, finasteride), male pattern baldness (finasteride), and female hirsutism (finasteride).

Women who are pregnant, or are trying to become pregnant, should avoid handling crushed or broken finasteride tablets.

dutasteride (Avodart) - capsule
finasteride (Proscar, Propecia) - tablet

Beta-Adrenergic Blocking Agents

By blocking the beta-adrenergic receptors, you can decrease the heart rate and the force of contractions causing a decrease in blood pressure. This means the drugs can treat hypertension, angina pectoris, tachycardia, and arrhythmias.

atenolol (Tenormin) - Tablet
bisoprolol (Zebeta) - Tablet
carvedilol (Coreg, Coreg CR) - extended-release capsule, tablet
labetalol (Trandate) - injectable solution, tablet
metoprolol (Lopressor, Toprol XL) - injectable solution, tablet, extended-release tablet
nebivolol (Bystolic) - tablet
propranolol (Inderal, Inderal LA) - oral solution, injectable solution, tablet, extended-release capsule
timolol (Timol, Timoptic, Timoptic XE) - ophthalmic solution, gel forming ophthalmic solution

Quick review question 9

If a patient were having a severe asthma attack, which medication might an emergency department administer?

a. norepinephrine
b. epinephrine
c. labetalol
d. doxazosin

Quick review question 10

Which of the following drugs are considered a beta-blocker?

a. Toprol XL
b. Tenex
c. Hytrin
d. Levophed

Drugs affecting the parasympathetic nervous system

The parasympathetic nervous system is part of the autonomic nervous system that tends to act in opposition to the sympathetic nervous system, by primarily regulating body functions during rest, digestion, and waste regulation. Stimulation of the parasympathetic system increases the activity of the gastrointestinal and genitourinary system while decreasing the activity of the cardiovascular system.

Cholinergic Drugs

The parasympathetic system is regulated by cholinergic receptors. The naturally occurring chemical that stimulates these receptors is called acetylcholine (ACH). Drugs that mimic ACH are therefore called cholinergic drugs. The conditions treated by this class of drugs varies widely. Donepezil and galantamine are each primarily used to treat Alzheimer's disease, while neostigmine is used to treat myasthenia gravis, and is an antidote for nondepolarizing neuromuscular blocking drugs (a group of drugs often used to create a neuromuscular blockade during surgery).

donepezil (Aricept, Aricept ODT) - tablet, orally disintegrating tablet
galantamine (Razadyne, Razadyne ER) - tablet, extended-release

tablet, oral solution
neostigmine (Prostigmin) - injectable solution, tablet

Anticholinergic Drugs

A naturally occurring chemical that works opposite of ACH is acetylcholinesterase, and therefore drugs that mimic acetylcholinesterase are often referred to as anticholinergic drugs. The conditions treated by this group of drugs also varies widely. Atropine can be used to increase the heart rate, anesthesia premedication, reversal of cholinergic drugs, treatment of GI spasticity, mydriasis, and enuresis treatment. Dicyclomine is typically used to treat GI disorders such as ulcers and colitis. Oxybutynin and tolterodine are typically used in the treatment of overactive bladder.

atropine (AtroPen, IsoptoAtropine) - injectable solution, tablet, ophthalmic solution, ophthalmic ointment
dicyclomine (Bentyl) - capsule, injectable solution, syrup, tablet
oxybutynin (Ditropan, Ditropan XL) - tablet, controlled-release tablet, syrup, transdermal patch, gel
tolterodine (Detrol, Detrol LA) -tablet, extended-release capsule

Quick review question 11

A patient, after surgery, is mostly likely to receive what medication to reverse the effects of vecuronium?

- a. neostigmine
- b. tolterodine
- c. atropine
- d. galantamine

Skeletal muscle relaxants

Skeletal muscle relaxants are used to treat conditions such as muscle spasticity and to relax muscle tone during surgeries. Medications that block muscle contraction within the spinal cord are referred to as centrally acting skeletal muscle relaxants; conversely peripherally acting skeletal muscle relaxants inhibit muscle contraction at the neuromuscular junction (NMJ).

All patients using skeletal muscle relaxants should avoid additional items that will depress the CNS or impair neuromuscular function such as alcohol, sedatives, and tranquilizers. There is also concern over abuse of these drugs, whether on their own or being used with other medications. Hence, carisoprodol was recently made a schedule IV controlled substance.

Neuromuscular Blocking Agents

Neuromuscular blocking agents (NMBAs or NMBs), also called peripherally acting skeletal muscle relaxants, are primarily used to prevent muscle contractions during surgeries or procedures where reflexes need to be suppressed (i.e., intubation). They can further be broken into two major subgroups: depolarizing (succinylcholine) and nondepolarizing (cisatracurium, pancuronium, rocuronium, and vecuronium).

cisatracurium (Nimbex) - injectable solution
pancuronium (Pavulon) - injectable solution
rocuronium (Zemuron) - injectable solution
succinylcholine, SUX (Anectine) - injectable solution
vecuronium (Norcuron) - powder for injection

Centrally Acting Skeletal Muscle Relaxants

Centrally acting skeletal muscle relaxants, also called spasmolytics, are primarily used to treat muscle spasms that may be caused by overexertion, trauma, or nervous tension. Baclofen and tizanidine are also used to treat multiple sclerosis.

baclofen (Lioresal) - tablet, injectable solution
carisoprodol (Soma) CIV - tablet
cyclobenzaprine (Flexeril) - tablet, extended-release capsule
metaxalone (Skelaxin) - tablet
methocarbamol (Robaxin) - tablet, injectable solution
tizanidine (Zanaflex) - tablet, capsule

Quick review question 12

A patient taking Flexeril due to spasms related to a traumatic back injury should avoid what?

  a. alcohol
  b. sedatives
  c. tranquilizers
  d. all of the above

Anesthetics

An anesthetic is a drug that produces anesthesia, a reversible loss of sensation.

Local Anesthetics

Local anesthetics, as their name implies, causes a temporary loss of feeling in a confined area of the body. Local anesthetics can be broken into two major groups: amide local anesthetics and ester local anesthetics. Cocaine is also used as a local anesthetic because despite its abuse potential, it is the only local anesthetic that causes vasoconstriction.

Ester Local Anesthetics

In general, ester local anesthetics have a short or moderate duration of action.

benzocaine (Solarcaine) - gel, topical solution, otic solution
cocaine CII - topical solution
tetracaine (Pontocaine) - injectable solution, powder for injection, throat spray, ophthalmic solution

Amide Local Anesthetics

Typically, amide local anesthetics have a long duration of action.

bupivacaine (Marcaine, Sensorcaine) - injectable solution
lidocaine (Xylocaine, Lidoderm) - injectable solution, infusion solution, oral solution, topical ointment, topical solution, topical jelly, patch

General Anesthetics

While many general anesthetics are available, the most common one for institutional pharmacies to dispense is propofol which is often used used for induction and maintenance of general anesthesia and ICU sedation for intubated, mechanically ventilated patients.

propofol (Diprivan) - injectable solution

Quick review question 13

Which local anesthetic causes vasoconstriction?

a. benzocaine
b. cocaine
c. bupivacaine
d. lidocaine

Medications for treating psychological conditions

Antipsychotic Drugs

Antipsychotic drugs, referred to as neuroleptics, are used to suppress the symptoms of schizophrenia and other psychotic conditions. Antipsychotics are associated with a range of side effects. Approximately two-thirds of patients will discontinue use due in part to adverse effects. Side effects may include acute dystonias, akathisia, parkinsonism, tardive dyskinesia, tachycardia, hypotension, impotence, lethargy, seizures, intense dreams or nightmares, and hyperprolactinaemia. Side effects from antipsychotics can be managed by a number of different drugs. For example, anticholinergics are often used to alleviate the motor side effects of antipsychotics. Some of the side effects will appear after the drug has been used only for a long time. When discontinuing therapy, patients must be tapered off of these drugs.

Clozapine is of particular concern because it can cause agranulocytosis and myocarditis. Therefore, patients receiving this medication need to be enrolled in a national registry and be closely monitored.

clozapine (Clozaril) - tablet, orally disintegrating tablet
haloperidol (Haldol) - tablet, oral concentrate, injectable solution, injectable solution-decanoate
olanzapine (Zyprexa, Zyprexa Zydis) - tablet, orally disintegrating tablet, short-acting IM injection, extended-release suspension IM injection
quetiapine (Seroquel, Seroquel XR) - tablet, extended-release tablet
risperidone (Risperdal, Risperdal Consta) - tablet, orally-disintegrating tablet, oral solution, powder for injection
ziprasidone (Geodon) - capsule, powder for injection

Hypnotics

Hypnotics are primarily used to induce and maintain sleep, usually to treat insomnia. There are three major categories of hypnotics: benzodiazepines, barbiturates, and nonbarbiturates. Other medications and substances that cause CNS depression (such as alcohol) should be avoided.

Benzodiazepines

Benzodiazepines are commonly used as anxiolytics, sedatives, hypnotics, anticonvulsants, and skeletal muscle relaxants.

alprazolam (Xanax) CIV - tablet, extended-release tablet, orally-disintegrating tablet, oral solution
clonazepam (Klonopin) CIV - tablet
diazepam (Valium) CIV - tablet, oral solution, rectal solution, injectable solution, intramuscular device
lorazepam (Ativan) CIV - tablet, oral concentrate, injectable solution
temazepam (Restoril) CIV - tablet

Barbiturates

Barbiturates are drugs that act as central nervous system depressants, and can therefore produce a wide spectrum of effects, from mild sedation to total anesthesia. They are also effective as anxiolytics, as hypnotics, and as anticonvulsants. Barbiturates also have analgesic effects; however, these effects are somewhat weak, preventing barbiturates from being used in surgery in the absence of other analgesics.

phenobarbital (Luminal) CIV - tablet, elixir, injectable solution

Nonbarbiturate Hypnotics

Barbiturates, while strictly considered hypnotics, have been largely replaced by the newer nonbarbiturates for treating insomnia since barbiturates are known to cause a 'hangover' effect.

eszopiclone (Lunesta) CIV - tablet
ramelteon (Rozerem) - tablet
zaleplon (Sonata) CIV - capsule
zolpidem (Ambien, Ambien CR) CIV - tablet, extended-release tablet, sublingual tablet, oral spray

Antidepressants

An antidepressant is a psychiatric medication used to alleviate mood disorders such as major depression and dysthymia, and anxiety disorders such as social anxiety disorder.

Antidepressants carry a black box warning that in short-term studies, antidepressants increased the risk of suicidal thinking and behavior in children, adolescents, and young adults.

Tricyclic Antidepressants

The tricyclic antidepressants are used primarily in the clinical treatment of mood disorders such as major depressive disorder, dysthymia, and treatment-resistant variants. They are also used in the treatment of a number of other medical disorders, including anxiety disorders such as generalized anxiety disorder, social phobia also known as social anxiety disorder, obsessive-compulsive disorder, and panic disorder, post-traumatic stress disorder, body dysmorphic disorder, eating disorders like anorexia nervosa and bulimia nervosa, certain personality disorders such as borderline personality disorder, attention-deficit hyperactivity disorder, as well as chronic pain, neuralgia or neuropathic pain, and fibromyalgia, headache, or migraine, smoking cessation, tourette syndrome, trichotillomania, irritable bowel syndrome, interstitial cystitis, nocturnal enuresis, narcolepsy, insomnia, pathological crying and/or laughing, chronic hiccups, ciguatera poisoning, and as an adjunct in schizophrenia.

amitriptyline (Elavil) - tablet
doxepin (Sinequan) - capsule, tablet, oral concentrate

Selective Serotonin Reuptake Inhibitors

Selective Serotonin Reuptake Inhibitors (SSRIs) are believed to work by decreasing the central nervous system's neuronal uptake of serotonin (5-HT). While these drugs can be used for a very broad range of things, you will typically see them used for treating moderate to major depression, generalized anxiety disorder, panic disorders, social anxiety, post-traumatic stress disorder, premenstrual dysphoric disorder, and prevention of migraine.

citalopram (Celexa) - tablet, oral solution
escitalopram (Lexapro) - tablet, oral solution
fluoxetine (Prozac) - tablet, capsule, delayed-release capsule, oral solution

fluvoxamine (Luvox) - tablet, extended-release capsule
paroxetine (Paxil) - tablet, extended-release tablet, oral suspension

Serotonin-Noerpinephrine Reuptake Inhibitors

Serotonin-norepinephrine reuptake inhibitors (SNRIs) prevent the uptake of neuronal serotonin and norepinephrine and are a less potent inhibitor of dopamine reuptake. The increase in these chemicals is believed to be related to their use in treating depression and anxiety. Duloxetine is also commonly used to treat chronic musculoskeletal pain, diabetic peripheral neuropathic pain, and fibromyalgia.

desvenlafaxine (Pristiq) - extended-release tablet
duloxetine (Cymbalta) - capsule
venlafaxine (Effexor, Effexor XR) - tablet, extended-release tablet, extended-release capsule

Miscellaneous Antidepressants

The medications in this category have varied mechanisms of action, but all function as antidepressants. Bupropion is also used for seasonal affective disorder, and smoking cessation. Mirtazapine is also used for post-traumatic stress disorder (PTSD). Trazadone is used for a number of things including aggressive behavior, alcohol withdrawal, insomnia, and prevention of migraine.

bupropion (Wellbutrin, Zyban) - tablet, sustained-release tablet, extended-release tablet
mirtazapine (Remeron, Remeron SolTab) - tablet, orally-disintegrating tablet
trazadone (Desyrel, Desyrel Dividose) - tablet, extended-release tablet

Antianxiety Agent, Nonbenzodiazepine

While many anxiolytic agents are related to benzodiazepines, one popular alternative is buspirone. It has a high affinity for 5HT1 receptors and a moderate affinity for dopamine D2 receptors. Buspirone is considered to have very little abuse potential; so unlike benzodiazepines, it is not a controlled substance.

buspirone (BusPar) - tablet

Partial Nicotinic Receptor Antagonist

Varenicline is used as a smoking cessation aid without actually using any form of nicotine. There is a heightened caution with this drug about suicidal ideation (suicidal thoughts).

varenicline (Chantix) - tablet

Amphetamines

Amphetamines are commonly used to treat attention deficit hyperactivity disorder (ADHD) and narcolepsy. Their ability to improve focus/concentration and boost energy levels has made this class of drugs very desirable for abuse; therefore, all amphetamines are considered schedule II controlled substances. Side effects may consist of severe weight loss; also, dependence may develop during use of this drug. Amphetamines can also raise the heart rate to dangerous levels.

amphetamine & dextroamphetamine (Adderall, Adderall XR) CII - tablet, capsule, extended-release capsule
dexmethylphenidate (Focalin, Focalin XR) CII - tablet, extended-release capsule
lisdexamfetamine (Vyvanse) CII - capsule
methylphenidate (Ritalin, Ritalin LA, Ritalin SR, Concerta, Daytrana) CII - tablet, chewable tablet, extended-release tablet, capsule, extended-release capsule, oral solution, transdermal patch

Anorexiant

Phentermine is indicated as a short-term (a few weeks) adjunct therapy for weight reduction based on increased exercise, behavior modification, and calorie reduction.

phentermine (Adipex P) CIV - tablet, orally-disintegrating tablet, capsule

Stimulant

Modafinil may increase dopamine in the brain by decreasing dopamine reuptake. Modafinil is used for treating narcolepsy, shift work sleep disorder, Obstructive Sleep Apnea/Hypopnea Syndrome (OSAHS), and off label it is used for fatigue in MS patients and for depression.

modafinil (Provigil) CIV - tablet

Antimanic Drug

Lithium is one of the oldest medications on the market for treating mania and is still quite popular. Lithium is also sometimes used to treat Huntington's disease, neutropenia (due to chemotherapy, or AIDS), cluster headache, PMS, bulimia, alcoholism, syndrome of inappropriate antidiuretic hormone secretion (SIADH), tardive dyskinesia, hyperthyroidism, and psychosis (postpartum or steroid-induced).

Lithium works by altering cation transport in nerve and muscle cells, and influences serotonin and/or norepinephrine reuptake.

Many of the antiepileptic drugs discussed later in this chapter are also often used as mood-stabilizing drugs.

lithium (Eskalith, Lithobid) - tablet, extended-release tablet, capsule, syrup

NMDA Antagonist

N-methyl-D-aspartate (NMDA) antagonists prevent excessive stimulation of the NMDA-receptor, which is ordinarily stimulated by glutamate. At normal levels, glutamate aids in memory and learning, but if levels are too high, glutamate appears to overstimulate nerve cells, killing them through excitotoxicity.

Memantine has been associated with a moderate decrease in clinical deterioration with only a small positive effect on cognition, mood, behavior, and the ability to perform daily activities in moderate to severe Alzheimer's disease. There does not appear to be any benefit in mild disease.

memantine (Namenda) - tablet, extended-release capsule, oral solution

Other Psychiatry Agents

Atomoxetine is used to treat ADHD. It functions as a selective inhibitor of presynaptic norepinephrine transport. Unlike amphetamines used to treat ADHD, atomoxetine is considered to have little or no abuse potential.

atomoxetine (Strattera) - capsule

Quick review question 14

Antidepressants carry a black box warning that in short-term studies:

  a. they may significantly reduce a patient's libido.
  b. these medications have been associated with nausea, vomiting, and diarrhea.
  c. antidepressants increased the risk of suicidal thinking and behavior in children, adolescents, and young adults.
  d. none of the above

Quick review question 15

All of the following classifications of drugs are frequently used as antidepressants except:

  a. selective serotonin reuptake inhibitors
  b. serotonin-norepinephrine reuptake inhibitors
  c. tricyclic antidepressants
  d. benzodiazepines

Quick review question 16

Which of the following drugs might be used for treating schizophrenia?

  a. olanzapine
  b. clonazepam
  c. zaleplon
  d. varenicline

Quick review question 17

Which of the following hypnotics is a schedule IV controlled substance?

  a. Xanax
  b. Luminal
  c. Ambien
  d. all of the above

Antiepileptic drugs

Antiepileptic drugs (AEDs) will also sometimes be referred to as anticonvulsants. All of the drugs in this category can be used for treating various types of seizures, and many of these medications have additional uses such as treating trigeminal neuralgia (carbamazepine), bipolar disorder (carbamazepine, lamotrigine, valproic acid), postherpetic neuralgia (gabapentin, pregabalin), muscle cramps (gabapentin), anxiety (gabapentin), diabetic neuropathy (gabapentin, pregabalin), fibromyalgia (pregabalin), neuropathic pain with spinal cord injury (pregabalin), migraine prophylaxis (topiramate, valproic acid), cluster-headache prophylaxis (topiramate), and alcoholism (topiramate).

carbamazepine (Tegretol, Equetro) - tablet, extended-release tablet, extended-release capsule, oral suspension
gabapentin (Neurontin) - capsule, tablet, oral solution
lamotrigine (Lamictal) - tablet, chewable tablet, orally-disintegrating tablet, extended-release tablet
levetiracetam (Keppra) - tablet, extended-release tablet, oral solution, injectable solution
oxcarbazepine (Trileptal) - tablet, extended-release tablet, oral suspension
pregabalin (Lyrica) CV - capsule, oral solution
topiramate (Topamax) - tablet, capsule
valproic acid (Depakote, Depakene, Depacon) - tablet, delayed-release tablet, extended-release tablet, capsule, delayed-release capsule, sprinkle capsule, syrup, injectable solution

Quick review question 18

Which of the following drugs may be used for treating seizures?

   a. carbamazepine
   b. gabapentin
   c. topiramate
   d. all of the above

Antiparkinson drugs

Parkinson's disease is a common neurologic disorder affecting approximately 1% of the population over the age of 60 (a much smaller percentage of the population will sometimes have this disease at an earlier age) caused by the loss of dopamine receptors. This disease presents a number of motor and nonmotor symptoms related to the loss of dopamine receptors.

Typically, the first motor symptom is a resting tremor in an upper extremity. Over time, additional motor symptoms occur including bradykinesia, rigidity, and gait difficulty. The first affected arm may not swing fully when walking, and the foot on the same side may scrape the floor. As the disease progresses, posture becomes increasingly flexed and strides become shorter causing a shuffling motion from the patient.

Nonmotor symptoms usually begin with the loss of smell, and is followed by rapid eye movements and behavior disorders.

The most common treatment is the use of carbidopa & levodopa to provide dopamine replacement therapy. Dopamine agonists (pramipexole, ropinirole) are also useful in decreasing the symptoms.

Dopamine agonists (pramipexole, ropinirole) are also useful in the treatment of restless leg syndrome.

carbidopa & levodopa (Sinemet, Sinemet CR, Parcopa) - tablet, orally-disintegrating tablet, extended-release tablet
pramipexole (Mirapex, Mirapex ER) - tablet, extended-release tablet
ropinirole (Requip, Requip XL) - tablet, extended-release tablet

Quick review question 19

Parkinson's disease is associated with the loss of which receptors?

a. alpha
b. beta
c. dopamine
d. acetylcholine

Analgesics

Opioid Analgesics

Opioid analgesics are often used to treat acute pain (such as post-operative pain), and for palliative care to alleviate the severe chronic, disabling pain of terminal conditions such as cancer, and sometimes for degenerative conditions such as rheumatoid arthritis. There has been an increased use of opioids in the management of non-malignant chronic pain. Opioids also have antitussive and antidiarrheal effects; therefore, low doses and or weaker forms of these medications will sometimes be used for those purposes as well. These medications do have a potential for abuse due to their addictive nature. While addictive themselves, two of these medications (buprenorphine & naloxine and methadone) can be used to treat opioid addiction. Most of these medications are considered controlled substances and their federal schedules are listed beside them.

buprenorphine & naloxone (Suboxone) CIII - sublingual tablet, sublingual film
codeine CII - tablet, oral solution
codeine & acetaminophen (Tylenol with Codeine, Tylenol #3, Tylenol #4) CIII as a tablet and C5 as an oral solution - tablet, oral solution
fentanyl (Duragesic, Sublimaze, Actiq) CII - transdermal patch, injectable solution, sublingual tablet, buccal tablet, buccal film, sublingual solution, lollipop
hydrocodone & acetaminophen (Vicodin, Lorcet) CIII - capsule, tablet, oral solution, oral elixir
hydromorphone (Dilaudid) CII - tablet, extended-release tablet, injectable solution, oral solution, suppository, powder for injection
meperidine (Demerol) CII - tablet, injectable solution, syrup
methadone (Methadose, Dolophine) CII - tablet, dispersible tablet, oral solution, injectable solution
morphine (MS Contin, Duramorph, Kadian) CII - tablet, extended-release capsule, controlled-release tablet, oral solution, injectable solution, suppository
oxycodone (OxyContin, Roxicodone) CII - tablet, capsule, extended-release tablet, oral solution
oxycodone & acetaminophen (Percocet, Tylox, Roxicet) CII - tablet, capsule, oral solution
tramadol (Ultram) not federally scheduled - tablet, orally-disintegrating tablet, extended-release tablet, extended-release capsule
tramadol & acetaminophen (Ultracet) not federally scheduled - tablet

Non Steroidal Anti Inflammatory Drugs

Nonsteroidal anti-inflammatory drugs, usually abbreviated to NSAIDs, are a class of drugs that provide analgesic and antipyretic effects, and, in higher doses, anti-inflammatory effects. NSAIDs are usually indicated for the treatment of acute or chronic conditions where pain and inflammation are present, such as rheumatoid arthritis, osteoarthritis, ankylosing spondylitis, gout, dysmenorrhoea, headache and migraine, postoperative inflammation, muscle stiffness, and fever.

celecoxib (Celebrex) - capsule
diclofenac (Voltaren, Cataflam) - tablet, extended-release tablet, capsule, powder packet for oral solution
ibuprofen (Advil, Motrin) - tablet, chewable tablet, oral suspension, injection solution
ketorolac (Toradol) - tablet, injectable solution
meloxicam (Mobic) - tablet, oral suspension
nabumetone (Relafen) - tablet
naproxen (Aleve, Naprosyn, Anaprox, Naprelan) - tablet, oral suspension
piroxicam (Feldene) - capsule

Miscellaneous Nonopioid Analgesics

Some medications do not easily fit into other categories, such as acetaminophen, aspirin and some of the combination analgesics that contain weak barbiturates.

Acetaminophen, sometimes listed as APAP, is a potent analgesic and antipyretic activity with weak anti-inflammatory activity. Acetaminophen-containing products pose a potential to harm liver function; therefore, patients are not to exceed a cumulative dose of 4 g/day of acetaminophen. In January 2011, the FDA mandated that all manufacturers limit acetaminophen in prescription products to 325 mg/dosage unit; manufacturers have until January 14, 2014, to comply.

Aspirin, sometimes listed as ASA, inhibits the synthesis of prostaglandin by cyclooxygenase; inhibits platelet aggregation; and has antipyretic and analgesic activity. While aspirin could technically be classified as a NSAID, it is more often classified as a salicylate.

One thing that surprises many about the combination of butalbital, acetaminophen, and caffeine is that it is not federally scheduled even though the combination of butalbital, aspirin, and caffeine is considered a controlled substance.

acetaminophen, APAP (Tylenol) - tablet, extended-release tablet, chewable tablet, disintegrating tablet, elixir, suspension, suppository
aspirin, ASA (Bayer, Ascriptin, Bufferin, Ecotrin, St. Joseph Adult Chewable Aspirin) - tablet, delayed-release tablet, effervescent tablet, chewable tablet, suppository
butalbital, acetaminophen, & caffeine (Fioricet) - tablet, capsule, oral solution
butalbital, aspirin, & caffeine (Fiorinal) CIII - tablet, capsule

Serotonin 5-HT Receptor Agonist

Serotonin 5-HT receptor agonists cause vasoconstriction in cranial arteries to relieve migraines and cluster headaches.

rizatriptan (Maxalt) - tablet, disintegrating tablet
sumatriptan (Imitrex) - tablet, nasal spray, injectable solution
zolmitriptan (Zomig) - tablet, disintegrating tablet, nasal spray

Quick review question 20

Which of the following opioid pain relievers is also used to treat opioid addiction?

a. meperidine
b. buprenorphine & naloxone
c. Vicodin
d. tramadol

Quick review question 21

Which of the following NSAIDs is available as an injectable solution?

a. celecoxib
b. diclofenac
c. ketorolac
d. meloxicam

Quick review question 22

Serotonin 5-HT agonists are primarily indicated for which of the following conditions?

a. migraines
b. fever
c. platelet aggregation

45

d. none of the above

Medications affecting the cardiac system

Cardiac Glycosides

Cardiac glycosides are drugs used in the treatment of congestive heart failure and cardiac arrhythmia.

digoxin (Lanoxin) - elixir, tablet, injectable solution

Diuretics

A diuretic elevates the rate of urination which removes fluid from the body and has a net result of decreased blood pressure. There are various mechanisms of action for these diuretics. Loop diuretics, such as bumetanide and furosemide, inhibit the reabsorption of sodium and chloride ions in the ascending loop of Henle within the kidneys' nephrons. By decreasing this reabsorption, an increased amount of fluid is excreted. Thiazide diuretics, such as chlorothiazide and hydrochlorothiazide (often abbreviated HCTZ), inhibit sodium reabsorption in distal renal tubules resulting in increased excrertion of sodium and water. Loop diuretics and thiazide diuretics also cause patients to lose a lot of potassium. Potassium sparing diuretics, such as spironolactone and triamterene, have an effect on renal distal tubules to inhibit sodium reabsorption causing the excretion of sodium and water but allows for the retention of potassium.

bumetanide (Bumex) - tablet, injectable solution
chlorothiazide (Diuril) - tablet, powder for injection
furosemide (Lasix) - tablet, oral solution, injectable solution
hydrochlorothiazide, HCTZ (Microzide) - tablet, capsule
spironolactone (Aldactone) - tablet
triamterene & hydrochlorothiazide (Dyazide, Maxzide) - tablet, capsule

Antiarrhythmics

Antiarrhythmic agents are a group of pharmaceuticals that are used to suppress abnormal rhythms of the heart (cardiac arrhythmias), such as atrial fibrillation, atrial flutter, ventricular tachycardia, and ventricular fibrillation.

amiodarone (Corddarone) - tablet, injectable solution
dronedarone (Multaq) - tablet

Angiotensin-II Receptor Antagonists

Angiotensin-II receptor antagonists (AIIRAs), also called angiotensin receptor blockers (ARBs), are a group of pharmaceuticals which modulate the renin-angiotensin-aldosterone system. Their main uses are in the treatment of hypertension (high blood pressure), diabetic nephropathy (kidney damage due to diabetes), and congestive heart failure.

candesartan (Atacand) - tablet
irbesartan (Avapro) - tablet
losartan (Cozaar) - tablet
olmesartan (Benicar) - tablet
telmisartan (Micardis) - tablet
valsartan (Diovan) - tablet

Angiotensin-Converting Enzyme Inhibitors

Angiotensin-converting enzyme inhibitors (ACE inhibitors) causes dilation of blood vessels which results in lower blood pressure. In treating heart disease, ACE inhibitors are usually used with other medications. A typical treatment plan will often include an ACE inhibitor, beta blocker, a long acting nitrate, and a calcium channel blocker in combinations that are adjusted to the individual patient's needs.

benazepril (Lotensin) - tablet
captopril (Capoten) - tablet
enalapril/enalaprilat (Vasotec) - tablet, injectable solution
fosinopril (Monopril) - tablet
lisinopril (Prinivil, Zestril) - tablet
quinapril (Accupril) - tablet
ramipril (Altace) - capsule

Calcium Channel Blockers

These drugs are used to treat hypertension. Calcium channel blockers work by inhibiting the influx of calcium ions into myocardial and vascular tissues thereby preventing contractions and causing the dilation of the main coronary and systemic arteries.

Patients should avoid grapefruit and its constituents while taking calcium channel blockers. Grapefruit inhibits an isoenzyme called CYP3A4 that helps breakdown calcium channel blockers. Inhibiting this enzyme causes patients to effectively be overdosed by their calcium channel blockers. Interactions related to grapefruit are often referred to as grapefruit juice drug interactions (GJDI).

amlodipine (Norvasc) - tablet
diltiazem (Cardizem, Cardizem CD, Taztia XT, Tiazac) - tablet, extended-release tablet, extended-release capsule, injectable solution, powder for injection
felodipine (Plendil) - extended-release tablet
nifedipine (Procardia, Procardia XL, Adalat CC) - capsule, extended-release tablet
verapamil (Isoptin, Calan, Covera HS) - tablet, extended-release tablet, extended-release capsule, injectable solution

Nitrates

Nitrates relax smooth muscle via dose-dependent dilation of arterial and venous beds to reduce both preload and afterload, and myocardial oxygen demand. Nitrates also improve coronary collateral circulation, lowers blood pressure, and increases heart rate. Nitrates are used for both the relief and prevention of angina pectoris, treatment of perioperative hypertension, control of congestive heart failure during a myocardial infarction and you will see these used off label for the treatment of anal fissures. Typically, patients need a nitrate free interval to minimize tolerance. This nitrate free interval is often provided at night.

Nitroglycerin also has special storage requirements due to its relative reactivity with certain plastics. As a result, the sublingual tablets should be stored in glass; typically, this means that patients should keep them in their original container from the manufacturer. Also, the nitroglycerin intravenous solution is infused with special tubing to prevent the drug from leaching into the plastic of most infusion sets.

isosorbide (Isordil, Dilatrate-SR, Imdur, ISMO, Monoket) - tablet, sublingual tablet, extended-release tablet, extended-release capsule
nitroglycerin, NTG (Nitrostat, Nitrol, NitroDur, Nitro-Bid, Nitrolingual Pumpspray) - sublingual tablet, extended-release capsule,

translingual solution, transdermal patch, intravenous solution, transdermal ointment, rectal ointment

Renin Inhibitor

A renin inhibitor prevents the conversion of angiotensinogen to angiotensin I. The decrease in antiotensin I causes a decrease in angiotensin II, a potent blood pressure elevating peptide. This makes renin inhibitors effective in the treatment of hypertension.

aliskiren (Tekturna) - tablet

Combination Antihypertensive Agents

Many antihypertensive medications are provided in combinations in order to combine effects and improve patient compliance by decreasing the number of medications the patient may need to take. As all the medications listed below have already been discussed in this chapter, you may refer to the previous information concerning each active ingredient within the combination medications listed below.

aliskiren & hydrochlorothiazide (Tekturna HCT) - tablet
amlodipine & benazepril (Lotrel) - capsule
amlodipine & olmesartan (Azor) - tablet
amlodipine & valsartan (Exforge) - tablet
bisoprolol & hydrochlorothiazide (Ziac) - tablet
enalapril & hydrochlorothiazide (Vaseretic) - tablet
irbesartan & hydrochlorothiazide (Avalide) - tablet
lisinopril & hydrochlorothiazide (Prinzide, Zestoretic) - tablet
losartan & hydrochlorothiazide (Hyzaar) - tablet
olmesartan & hydrochlorothiazide (Benicar HCT) - tablet
telmisartan & hydrochlorothiazide (Micardis HCT) - tablet
valsartan & hydrochlorothiazide (Diovan HCT) - tablet

Anticoagulants and Antiplatelet Agents

A broad range of chemicals with various mechanisms of action fall into this category. Anticoagulants and antiplatelet drugs are used for the following treatments: prophylaxis after a myocardial infarction or stroke (aspirin & dipyridamole, clopidogrel, warfarin), acute coronary syndrome (clopidogrel, heparin), peripheral arterial disease (clopidogrel), coronary artery disease (clopidogrel), stenting (clopidogrel), atrial fibrilation (dabigatran, rivaroxaban, warfarin), thromboembolism (dabigatran, dalteparin), deep vein thrombosis (dalteparin, enoxaparin, heparin, rivaroxaban, warfarin), pulmonary embolism (heparin, rivaroxaban, warfarin), unstable angina (dalteparin, enoxaparin, heparin), anticoagulation therapy (dalteparin, heparin), catheter patency (heparin), cardiac valve replacement (warfarin).

Medication interactions are a common concern with these drugs as many pain relievers (most NSAIDs and aspirin) may slow a patient's clotting time. These pain relievers should not be taken in combination with the various anticoagualant and antiplatelet drugs without first consulting a physician.

Warfarin also provides a significant food-drug interaction. Warfarin prevents coagulation by binding vitamin K; therefore foods rich in vitamin K (such as dark leafy greens, broccoli, and asparagus) should be avoided as they will interfere with warfarin.

Below is a short list of the most common anticoagulants and antiplatelet agents.

aspirin & dipyridamole (Aggrenox) - extended-release capsule
clopidogrel (Plavix) - tablet
dabigatran (Pradaxa) - capsule
dalteparin (Fragmin) - injectable solution
enoxaparin (Lovenox) - injectable solution
heparin - heparin lock solution, injectable solution, premixed IV solution
rivaroxaban (Xarelto) - tablet
warfarin - (Coumadin) - tablet, powder for injection

HMG-CoA Reductase Inhibitors

HMG-CoA reductase inhibitors (commonly referred to as statins) block the pathway for synthesizing cholesterol in the liver. This is significant because most circulating cholesterol comes from internal manufacture rather than the diet. When the liver can no longer produce cholesterol, total levels of cholesterol in the blood will fall, particularly LDL-cholesterol and triglycerides. They also increase "good cholesterol" -- HDL cholesterol. Cholesterol synthesis appears to occur mostly at night, so statins with short half-lives are usually taken at night to maximize their effect. The most common adverse side effects are raised liver enzymes and muscle problems (most frequently muscle cramps).

Patients should avoid grapefruit and its constituents while taking HMG-CoA reductase inhibitors. Grapefruit inhibits an isoenzyme called CYP3A4 that helps breakdown HMG-CoA reductase inhibitors. Inhibiting this enzyme causes patients to effectively be overdosed by their HMG-CoA reductase inhibitors. Interactions related to grapefruit are often referred to as grapefruit juice drug interactions (GJDI).

atorvastatin (Lipitor) - tablet
fluvastatin (Lescol, Lescol XL) - capsule, extended-release tablet
lovastatin (Mevacor, Altoprev) - tablet, extended-release tablet
pravastatin (Pravachol) - tablet
rosuvastatin (Crestor) - tablet
simvastatin (Zocor) - tablet

Miscellaneous Hypolipidemic Drugs

There are a number of other medications used to reduce cholesterol with various mechanisms of action. Amlodipine & atorvastatin (Caduet) are used in combination to treat both hypertension and reduce cholesterol.

amlodipine & atorvastatin (Caduet) - tablet
ezetimibe (Zetia) - tablet
fenofibrate (Tricor) - tablet, capsule
gemfibrozil (Lopid) - tablet
niacin, vitamin B3, nicotinic acid (Niacor, Niaspan) - tablet, extended-release tablet, extended-release capsule
omega 3 fatty acids (Lovaza) - chewable tablet, capsule, delayed-release capsule
simvastatin & ezetimibe (Vytorin) - tablet

Quick review question 23

Which of the following is considered a potassium sparing diuretic?

  a. Lasix
  b. HCTZ
  c. Aldactone
  d. Tikosyn

Quick review question 24

Which of the following is the generic name for Avapro?

a. irbesartan
b. losartan
c. telmisartan
d. valsartan

Quick review question 25

Which of the following antihypertensive medications is considered a calcium channel blocker?

a. Capoten
b. Cardizem
c. Coreg
d. Dilatrate-SR

Quick review question 26

Which of the anticoagulants listed below are available in an oral formulation?

a. dalteparin
b. enoxaparin
c. heparin
d. warfarin

Quick review question 27

Atorvastatin is used to:

a. increase triglyceride levels
b. increase LDL cholesterol
c. decrease HDL cholesterol
d. decrease total levels of cholesterol

Antihistamines

Colds and allergy sufferers may desire symptomatic relief from antihistamine and decongestants.

Antihistamines

Antihistamines are commonly used for the relief of various allergic reactions including allergic rhinitis, perennial & seasonal allergies, pruritus, and allergic conjunctivitis. First generation antihistamines (diphenhydramine and hydroxyzine) can cross the blood brain barrier. As a result, diphenhydramine may be used to treat insomnia, and hydroxyzine may be used to treat nausea and vomiting.

cetririzine (Zyrtec) - tablet, chewable tablet, capsule, oral solution
desloratadine (Clarinex, Clarinex RediTabs) - tablet, disintegrating tablet, syrup
diphenhydramine (Benadryl) - tablet, chewable tablet, disintegrating tablet, capsule, oral solution, cream, gel, ointment, lotion, topical aerosol spray, oral strip, injectable solution
fexofenadine (Allegra) - tablet, disintegrating tablet, oral suspension
hydroxyzine (Vistaril) - tablet, capsule, syrup, oral suspension, injectable solution
loratadine (Claritin) - tablet, chewable tablet, disintegrating tablet, capsule, oral solution
olopatadine (Patanol, Patanase) - ophthalmic drops, nasal spray

Antihistamines and Decongestants

Patients with either colds or congestion related to their allergies may desire an antihistamine in combination with a decongestant. Since pseudoephedrine is the most commonly included decongestant, many of the medications need to be kept behind the counter in order to limit their availability. This is a result of the Combat Methamphetamine Epidemic Act, which will be discussed further in the next chapter.

cetirizine & pseudoephedrine (Zyrtec D) - tablet, extended-release tablet
desloratadine & pseudoephedrine (Clarinex-D) - extended-release tablet
fexofenadine & pseudoephedrine (Allegra D) - extended-release tablet
loratadine & pseudoephedrine (Claritin D) - extended-release tablet

Quick review question 28

Which antihistamine is sometimes prescribed to treat insomnia?

a. cetirizine
b. diphenhydramine
c. fexofenadine

d. olopatadine

Respiratory medications

Some patients may have respiratory conditions such as asthma or bronchitis. Patients may receive glucocorticosteroids, bronchodialators, leukotriene receptor antagonists, or even epinephrine to treat or prevent acute episodes. Glucocorticosteroids and epinephrine have already been discussed in this chapter; therefore, this section will focus on bronchodialators and leukotriene receptor antagonists.

Bronchodialators

A bronchodialator is a substance that dilates the bronchi and bronchioles, decreasing resistance in the respiratory airway and increasing airflow to the lungs. The bronchodialators on this list can be further broken down into beta-2 agonists (albuterol, levalbuterol, and salmeterol) and anticholinergics (ipratropium and tiotropium). Fluticasone, which may be used on it own or in combination with other drugs, is classified as a glucocorticosteroid.

albuterol (Proventil HFA, Ventolin HFA, Proair HFA) - tablet, oral liquid, inhalation aerosol, solution for nebulization
albuterol & ipratropium (Combivent, DuoNeb) - inhalation aerosol, solution for nebulization
budesonide & formoterol (Symbicort) - inhalation aerosol
ipratropium (Atrovent) - inhalation aerosol, solution for nebulization, nasal spray
levalbuterol (Xopenex) - inhalation aerosol, solution for nebulization
salmeterol & fluticasone (Advair Diskus, Advair HFA) - disk with powder for inhalation, inhalation aerosol
tiotropium (Spiriva) - capsules for inhalation

Miscellaneous Asthma Treatments

Leukotriene receptor antagonists have been shown to improve asthma symptoms, reduce asthma exacerbations, and limit markers of inflammation. These medications are also used for allergies.

montelukast (Singulair) - tablet, chewable tablet, granules for suspension *zafirlukast* (Accolate) - tablet

Quick review question 29

Which of the following medications require a special device for patients to inhale capsules?

   a. Proventil
   b. Combivent
   c. Spiriva
   d. Singulair

Medications affecting the Gastrointestinal System

The gastrointestinal system stretches from mouth to anus and includes the esophagus, stomach, and intestines. The long portion of the body can have a plethora of ailments from heart burn and reflux, to stomach and duodenal ulcers, to emesis and diarrhea. This section endeavors to look at some of the more common medications utilized to cover these conditions.

H2-Receptor Antagonists

H2-receptor antagonists are used to block the action of histamine on parietal cells in the stomach, decreasing the production of acid by these cells. H2-antagonists are used for peptic ulcer disease (PUD), gastroesophageal reflux disease (GERD), dyspepsia, and the prevention of stress ulcers.

cimetidine (Tagamet) - tablet, oral solution
famotidine (Pepcid) - tablet, chewable tablet, oral suspension, injectable solution, premixed IV bag
nizatidine (Axid) - capsule, oral solution
ranitidine (Zantac) - tablet, chewable tablet, effervescent tablet, capsule, syrup, injectable solution, premixed IV bag

Proton Pump Inhibitors

Proton-pump inhibitors (PPIs) produce a pronounced and long-lasting reduction of gastric acid production. These drugs are used to treat dyspepsia, peptic ulcer disease (PUD), gastroesophageal reflux disease (GERD), laryngopharyngeal reflux, Barrett's esophagus, stress gastritis prevention, gastrinomas and other conditions that cause hypersecretion of acid, and Zollinger-Ellison syndrome.

dexlansoprazole (Dexilant) - delayed-release capsule
esomeprazole (Nexium) - delayed-release capsule, granules for suspension, injectable solution
lansoprazole (Prevacid) - disintegrating tablet, delayed-release capsule, oral suspension
omeprazole (Prilosec) - delayed-release tablet, delayed-release capsule, oral suspension, powder for oral suspension
pantoprazole (Protonix) - delayed-release tablet, powder for oral suspension, powder for injection
rabeprazole (Aciphex) - delayed-release tablet

Antiemetics

Antiemetics are used to treat nausea and vomiting associated with a number of things including motion sickness, the side effects of opioid analgesics and general anesthetics, and chemotherapy directed against cancer. Antiemetics are also used for morning sickness; however, there is little information about the effect on the fetus, and therefore are typically reserved for times considered strictly necessary.

aprepitant (Emend) - capsule
dolasetron (Anzemet) - tablet, injectable solution
dronabinol (Marinol) CIII - capsule
granisetron (Kytril) tablet, transdermal patch, injectable solution
meclizine (Antivert, Bonine) - tablet, chewable tablet
metoclopramide (Reglan) - tablet, dispersible tablet, oral solution, injectable solution
prochlorperazine (Compazine) - tablet, suppository, injectable solution
promethazine (Phenergan) - oral tablet, syrup, injectable solution, suppository
ondansetron (Zofran) - tablet, dispersible tablet, oral solution, oral film, injectable solution
scopolamine (Transderm Scop) - tablet, transdermal patch, ophthalmic solution, injectable solution
trimethobenzamide (Tigan) - capsule, intramuscular solution

Stool Softeners and Laxatives

Stool softeners and laxatives are taken to loosen stool and treat constipation. Some laxatives are used to evacuate the colon for rectal and/or bowel examinations.

bisacodyl (Dulcolax) - delayed-release tablet, suppository, enema
docusate (Colace) - tablet, capsule, syrup
docusate & senna (Peri-Colace, Senokot-S) - tablet
lactulose (Enulose, Kristalose) - oral solution, powder for oral solution
polyethylene glycol (MiraLax, Glycolax) - powder for oral solution
polyethylene glycol with electrolytes (GoLytely) - oral solution, powder for oral solution
psyllium (Metamucil) - powder for oral solution
senna (Senokot) - tablet, chewable tablet, syrup

Quick review question 30

Which of the following antiemetics is often used for motion sickness and is available as a patch ?

a. scopolamine
b. trimethobenzamide
c. ondansetron
d. dexlansoprazole

Quick review question 31

Which of the following is an H2-receptor antagonist?

a. Protonix
b. Prilosec
c. Prevacid
d. Pepcid

Drugs affecting the thyroid gland and bone degeneration

The thyroid gland secretes three hormones: triiodothyronine (T_3), thyroxine (T_4), and thyrocalcitonin. Triiodothyronine and thyroxine help regulate tissue growth and mitochondrial metabolism in most of the cells in the human body. Thyrocalcitonin primarily affects bone formation.

Thyroid Replacement Therapy

Thyroid replacement therapies are primarily used to treat hypothyroidism. These therapies use medications that replace T3 (liothyronine), T4 (levothyroxine), or both T3 and T4 (thyroid).

levothyroxine (Synthroid, Levoxyl, Levothroid) - tablet, capsule, powder for injection
liothyronine (Cytomel, Triostat) - tablet, injectable solution
thyroid (Armour Thyroid) - tablet

Osteoporosis Therapy

Osteoporosis and Paget's Disease are diseases in which bones have decreased mass and and are more fragile. While calcium and vitamin D (both discussed further when talking about dietary supplements in a later section) are important to developing and maintaining strong bones, other medications are used to specifically combat these diseases, such as calcitonin (used to supplement thyrocalcitonin), bisphosphonate derivatives (such as alendronate, ibandronate, risedronate, and zoledronic acid), and parathyroid hormone analog (teriparatide). Patients should be instructed to remain upright for 60 minutes after taking a bisphosphonate derivative to prevent stomach upset, and inflammation and erosion of the esophagus.

alendronate (Fosamax) - tablet, weekly tablet, effervescent tablet, oral solution
calcitonin (Miacalcin, Fortical) - injectable solution, nasal spray
ibandronate (Boniva) - monthly tablet, prefilled syringe
risedronate (Actonel) - tablet, weekly tablet
teriparatide (Forteo) - prefilled injectable pen
zoledronic acid (Reclast, Zometa) - injectable solution

Quick review question 32

Paget's Disease affects which of the following?

a. GI tract
b. mitochondrial metabolism
c. bones
d. liver

Quick review question 33

Which of the following drugs would not be used to treat hypothyroidism?

a. levothyroxine
b. lansoprazole
c. thyroid
d. liothyronine

Medications for treating diabetes mellitus

Diabetes mellitus refers to a group of diseases that affect how your body uses blood glucose, commonly called blood sugar. There are two primary types of this disease, type 1 diabetes and type 2 diabetes.

Type 1 diabetes, also known as insulin dependent diabetes mellitus (IDDM) or juvenile diabetes, typically appears during childhood or adolescence. Individuals with type 1 diabetes require insulin therapy as their bodies do not produce insulin.

Type 2 diabetes, also known as non-insulin dependent diabetes mellitus (NIDDM), usually occurs in adults. Individuals with type 2 diabetes either don't produce enough insulin or their cells have become insulin resistant. These patients will usually receive non-insulin therapies, although they may receive insulin therapy, or even a combination of insulin and non-insulin therapies.

Insulins

Insulin therapy is required in type 1 diabetes, and may be necessary in some individuals with type 2 diabetes. The general objective of insulin replacement therapy is to approximate the physiological pattern of insulin secretion. This requires a basal insulin throughout the day, supplemented by prandial insulin at mealtime. Insulin injections are intended to mimic the natural process shown in the following image.

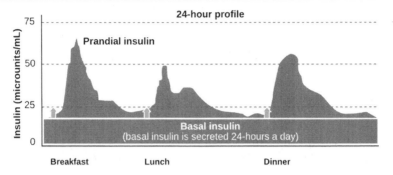

Conceptual depiction of insulin profiles

A combination of rapid-acting (insulin aspart, insulin glulisine, insulin lispro) or short-acting (insulin regular) given in combination with either an intermediate-acting (insulin NPH) or long-acting (insulin detemir, insulin glargine) insulin are typically used. Sometimes an insulin vial or prefilled syringe may have a mixture of insulins (insulin aspart protamine & insulin aspart, insulin lispro protamine & insulin lispro, insulin NPH & insulin regular human) to help reduce the number of injections a patient requires.

insulin aspart (NovoLog) - vial for injection, prefilled pen
insulin aspart protamine & insulin aspart (Novolog Mix 70/30) - vial for injection, prefilled pen
insulin detemir (Levemir) - vial for injection, prefilled pen
insulin glargine (Lantus) - vial for injection, prefilled pen
insulin glulisine (Apidra) - vial for injection, prefilled pen
insulin lispro (Humalog) - vial for injection, prefilled pen
insulin lispro protamine & insulin lispro (Humalog Mix 50/50, Humalog Mix 75/25) - vial for injection, prefilled pen
insulin NPH (Humulin N, Novolin N) - vial for injection, prefilled pen
insulin NPH & insulin regular human (Humulin 70/30, Novolin 70/30) - vial for injection, insulin pen
insulin regular human (Humulin R, Novolin R) - vial for injection, prefilled pen, concentrated vial for injection

Biguanide

Metformin decreases hepatic glucose production, decreases GI glucose absorption, and increases target cell insulin sensitivity.

60

metformin (Glucaphage, Glucophage XR) - tablet, extended-release tablet, oral solution

Sulfonylureas

Sulfonylureas initially increase insulin secretion, increase insulin receptor sensitivity, and may decrease liver production of glucose. Some patients have an increased risk of hypoglycemia from sulfonylureas. These medication have a cross-sensitivity with sulfa allergies.

glimepiride (Amaryl) - tablet
glipizide (Glucotrol, Glucotrol XL) - tablet, extended-release tablet
glyburide (Diabeta, Glynase, Micronase) - tablet, micronized tablet

Metaglitinide

Metaglitinides increase insulin secretion.

nateglinide (Starlix) - tablet
repaglinide (Prandin) - tablet

Thiazolidinediones

Thiazolidinediones improve cellular response to insulin and decreases the liver's production of glucose. Thiazolidinediones, particularly products containing rosiglitazone, may cause or exacerbate congestive heart failure in some patients.

pioglitazone (Actos) - tablet
rosiglitazone (Avandia) -tablet

Dipeptidyl Peptidase-4 Inhibitors

Dipeptidyl peptidase-4 (DPP-4) degrade incretin hormones. DPP-4 inhibitors increase and prolong incretin hormone activity. Incretins increase insulin release and synthesis.

linagliptin (Tradjenta) - tablet
saxagliptin (Onglyza) - tablet
sitagliptin (Januvia) - tablet

Glucagonlike Peptide-1 Agonists

Glucagonlike peptide-1 mimic incretin, causing an increase in insulin release and synthesis.

exenatide (Byetta) - injectable solution
liraglutide (Victoza) - injectable solution

Oral Antidiabetic Combination Drugs

Often, type 2 diabetics will require multiple medications. Combination products provide convenience and improved patient compliance.

glipizide & metformin (Metaglip) - tablet
glyburide & metformin (Glucovance) - tablet
sitagliptin & metformin (Janumet) - tablet

Quick review question 34

Which of the following medications would not be of benefit to a type 1 diabetic?

  a. Byetta
  b. Apidra
  c. Lantus
  d. Humulin R

Quick review question 35

A type 2 diabetic that has previously had an allergic reaction to Bactrim should not receive which of the following medications?

  a. NovoLog Mix 50/50
  b. Byetta
  c. Amaryl
  d. Janumet

Quick review question 36

Type 1 diabetic require insulin coverage to mimic which of the following?

  a. basal insulin
  b. prandial insulin
  c. both basal and prandial insulin
  d. none of the above

Quick review question 37

Metformin does which of the following?

a. decreases hepatic glucose production
b. decreases GI glucose production
c. increases target cell insulin sensitivity
d. all of the above

Antineoplastic agents and monoclonal antibodies

Antineoplastics and monoclonal antibodies are types of chemotherapy. Chemotherapy is the use of medications to kill cancer cells. Chemotherapy is used for diffuse tumors and is used after surgery and irradiation of solid tumors in the attempt to eliminate remaining cancer cells that have metastasized.

Antineoplastic Agents

Antineoplastic agents are intended to inhibit uncontrolled new cell growth. Antineoplastic agents are used to treat breast cancer (anastrazole, capecitabine, cyclophosphamide, docetaxel, doxorubicin, exemestane, fluorouracil, gemcitabine, letrozole, leuprolide, mitomycin, paclitaxel, tamoxifen), non-small cell lung cancer (cisplatin, docetaxel, gemcitabine, irinotecan, mitomycin, paclitaxel), small cell lung cancer (carboplatin, doxorubicin, etoposide, paclitaxel, vincristine), pleural sclerosing (bleomycin), brain cancer (paclitaxel, vincristine), head and neck cancer (bleomycin, carboplatin, cisplatin, docetaxel, doxorubicin, fluorouracil, mitomycin, paclitaxel, vinblastine), thyroid cancer (doxorubicin), Hodgkin's disease (bleomycin, etoposide, vinblastine, vincristine), Non-Hodgkin's disease (bleomycin, cyclophosphamide, docetaxel, etoposide, paclitaxel, vincristine), lymphoma (bleomycin, carboplatin, vinblastine, vincristine), leukemia (daunorubicin, etoposide), meningeal leukemia (methotrexate), lymphoblastic leukemia (methotrexate), osteosarcoma (cisplatin, methotrexate), multiple myeloma (mitomycin), esophageal cancer (cisplatin, fluorouracil, mitomycin, paclitaxel), stomach cancer (docetaxel, doxorubicin, fluorouracil, mitomycin, paclitaxel), colon cancer (capecitabine, fluorouracil, irinotecan, oxaliplatin), colorectal cancer (capecitabine, fluorouracil, irinotecan, oxaliplatin), rectal cancer (fluorouracil, irinotecan), bladder cancer (cisplatin, fluorouracil, gemcitabine, mitomycin, vinblastine, vincristine), urothelial cancer (docetaxel, paclitaxel), ovarian carcinoma (carboplatin, cisplatin, docetaxel, doxorubicin, fluorouracil, gemcitabine, irinotecan, letrozole, oxaliplatin, paclitaxel, vinblastine), cervical cancer (bleomycin, carboplatin, cisplatin, fluorouracil), vulvar cancer (bleomycin), endometrial cancer (cisplatin, fluorouracil), endometriosis (leuprolide), uterine fibroids (leuprolide), testicular carcinoma (bleomycin, carboplatin, cisplatin, etoposide, vinblastine), prostate cancer (cisplatin, docetaxel, doxorubicin, fluorouracil, leuprolide, paclitaxel, vinblastine), penile cancer (bleomycin), pancreatic cancer (bleomycin, fluorouracil, gemcitabine, irinotecan, mitomycin, paclitaxel), kidney cancer (cisplatin), liver cancer (doxorubicin, fluorouracil), melanoma (docetaxel, paclitaxel, vinblastine, vincristine), actinic keratoses (fluorouracil), superficial basal cell carcinoma (fluorouracil), Kaposi's sarcoma (etoposide, paclitaxel, vinblastine, vincristine), rhabdomyosarcoma (vincristine), neuroblastoma (vincristine), neoplasms (methotrexate), Wilm's tumor (etoposide, vincristine), soft tissue sarcoma (docetaxel), malignant neoplastic diseases (cyclophosphamide), psoriasis (methotrexate), nephrotic syndrome (cyclophosphamide), juvenile idiopathic arthritis (cyclophosphamide), rheumatoid arthritis, (methotrexate), lupus nephritis (cyclophosphamide), and systemic sclerosis (cyclophosphamide).

These drugs often work during specific phases of the cell cycle as depicted in the image.

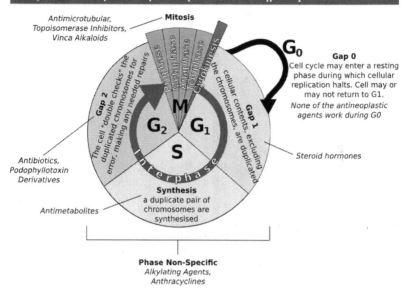

Cell cycle with classifications of antineoplastics used in different phases listed in italics

Unfortunately, antineoplastic agents can not differentiate between cancer cells and healthy cells which is responsible for many of the common side effects from these treatments as there is rapid cell growth in bone marrow, the GI tract, hair follicles, and skin cells. This often results in anemias, nausea, vomiting, diarrhea, hair loss, photosensitivity, and thinner skin. Many of these medications also have their own unique adverse effects. To learn more about these drugs go to: http://reference.medscape.com/drugs/oncology.

Alkylating Agents

These medications are considered phase non-selective for when they work; therefore, they work during all phases, except for the G_0 phase.

Besides the broad grouping of side effects common to antineoplastic agents, these drugs have the following side effects as well: carboplatin causes delayed thrombocytopenia; cisplatin has ototoxicity and can cause renal damage; cyclophosphamide can cause hemorrhagic cystitis; oxaliplatin has neurotoxicities, renal toxicity, and can trigger hypersensitivity reactions.

carboplatin (Paraplatin) - powder for injection, injectable solution
cisplatin (Platinol) - injectable solution
cyclophosphamide (Cytoxan) - powder for injection, tablet
oxaliplatin (Eloxatin) - powder for injection

Anthracyclines

These medications are considered phase non-selective for when they work; therefore, they work during all phases, except for the G_0 phase.

Besides the broad grouping of side effects common to antineoplastic agents, they can cause stomatitis, and have cardiotoxicities. These drugs are also considered vesicants.

daunorubicin (Cerubidine) - injectable solution, powder for injection
doxorubicin (Adriamycin) - injectable solution, powder for injection

Antibiotics

These drugs are G_2 phase specific.

These medications have pulmonary toxicities. Bleomycin may cause fever/chill and skin erythema. Mitomycin is also considered a vesicant.

bleomycin (Blenoxane) - powder for injection
mitomycin (Mitomycin C, Mutamycin) - powder for injection

Antimetabolites

These drugs are S phase specific.

All of these medications may cause stomatitis. Gemcitabine may also cause a flu-like syndrome. Methotrexate can cause renal dysfunction.

capecitabine (Xeloda) - tablet
fluorouracil, 5-FU (Adrucil, Efudex) - injectable solution, cream, topical solution
gemcitabine (Gemzar) - powder for injection
methotrexate, MTX (Trexall, Rheumatrex) - injectable solution, powder for injection, tablet

Antimicrotubular Drugs

These drugs are mitosis phase specific.

Additional side effects with these medications include peripheral neuropathy and hypersensitivity reactions.

docetaxel (Taxotere) - injectable solution
paclitaxel - injectable solution

Podophyllotoxin Derivatives

This drug is G_2 phase specific.

An additional side effect of etoposide is that it can cause or exacerbate hypotension.

etoposide (VePesid, vp 16) - capsule, injectable solution, powder for injection

Topoisomerase Inhibitors

This drug is mitosis phase specific.

irinotecan (Camptosar) - injectable solution

Vinca Alkaloids

These drugs are mitosis phase specific.

These drugs have neurotoxicities and they are both also considered vesicants.

vinblastine (Velban) - injectable solution, powder for injection
vincristine (Oncovin) - injectable solution

Hormone Antagonists

Tumors that involve the reproductive organs (breast, ovaries, uterus, prostate and testes) are often hormone-dependent. The hormones often responsible for developing and maintaining these organs are now functioning as growth factors for these cancerous cells. Hormone antagonists can be useful in the treatment of these cancers. Hormone antagonists tend to work best during the G_1 phase by pushing these cancer cells into the G_0 phase.

These medications are often used in combination with surgery and/or radiation therapy.

Many of the side effects from other chemotherapies are not as severe or are often absent.

Selective Estrogen Receptor Modulators

Tamoxifen is typically used in premenopausal women for 5-10 years.

tamoxifen (Nolvadex) - tablet, oral solution

Aromatase Inhibitors

anastrazole (Arimidex) - tablet
exemestane (Aromisin) - tablet
letrozole (Femara) - tablet

GnRH Agonists

leuprolide (Lupron) - solution for injection, prefilled syringe

Monoclonal antibodies

Monoclonal antibodies, often referred to as targeted therapy, only works against specific cells. This can reduce or even eliminate many of the side effects, but the medications tend to be very limited as to which cancers they can treat. This is because monoclonal antibodies are monospecific antibodies that are all identical, because they are made by immune cells that are all clones of a unique parent cell.

Monoclonal antibodies are used to treat glioblastoma (bevacizumab), age related macular degeneration (bevacizumab), breast cancer (bevacizumab, trastuzumab), non-small cell lung cancer (bevacizumab), peritoneal carcinoma (bevacizumab), leukemia (alemtuzumab, rituximab), Non-Hodgkin's disease (rituximab), renal cell carcinoma (bevacizumab), stomach cancer (bevacizumab), gastric cancer (trastuzumab), colorectal cancer (bevacizumab), fallopian tube carcinoma (bevacizumab), ovarian cancer (bevacizumab), pancreatic cancer (bevacizumab, trastuzumab), melanoma (bevacizumab), Wegener granulomatosis (rituximab), idiopathic pulmonary fibrosis (infliximab), respiratory syncytial virus (palivizumab), sarcoidosis (infliximab), rheumatoid arthritis (adalimumab, infliximab, rituximab), psoriatic arthritis (adalimumab, infliximab), ankylosing spondylitis (adalimumab, infliximab), Crohn's disease (adalimumab, infliximab), ulcerative colitis (adalimumab, infliximab), plaque psoriasis (adalimumab, infliximab), microscopic polyangiitis (rituximab), and immune thrombocytopenic purpura (rituximab).

adalimumab (Humira) - prefilled pen
alemtuzumab (CamPath) - injectable solution
bevacizumab (Avastin) - injectable solution
infliximab (Remicade) - powder for injection
palivizumab (Synagis) - injectable solution
rituximab (Rituxan) - injectable solution
trastuzumab (Herceptin) - powder for injection

Quick review question 38

A common side effect of chemotherapy often include which of the following?

a. anemia
b. nausea and vomiting
c. hair loss
d. all of the above

Quick review question 39

Why are monoclonal antibodies often referred to as targeted therapy?

a. they only work against specific cells
b. mice are often used to create the hybridomas
c. they are derived from dead animals that have been carefully hunted
d. they are not referred to as targeted therapy

Miscellaneous drug categories

The remaining medications in this list are difficult to group together with other drugs. The following is a short list of unrelated categories.

Hemopoietics

These medications affect the hemopoietic stem cells in bone marrow causing an increase in either red blood cells (darbepoetin, epoetin), white blood cells (filgrastim), or platelets (oprelvekin). Dialysis, chemotherapy, and various other causes of anemia make this class of medications extremely useful.

darbepoetin alfa (Aranesp) - injectable solution, prefilled syringe
epoetin alfa (Epogen, Procrit) - injectable solution
filgrastim (Neupogen) - injectable solution, prefilled syringe
oprelvekin (Neumega) - powder for injection

Gouty Arthritis Treatment

Gouty arthritis is painful inflammation (often of the big toe and foot) resulting from elevated levels of uric acid in the blood and the deposition of urate crystals around the joints. This disease has a greater occurrence in men. The condition can become chronic and result in deformity.

Allopurinol, azathioprine, and colchicine can all be used to treat gouty arthritis. Allopurinol is also used to treat antineoplastic induced hyperuricemia. Azathioprine can also be used for prevention of kidney transplant rejection, rheumatoid arthritis, lupus nephritis, Behcet's syndrome, Crohn's disease, myasthenia gravis, and multiple sclerosis. Colchicine can also be used for familial Mediterranean fever, and Behcet's syndrome.

allopurinol (Zyloprim) - tablet, powder for injection
azathioprine (Imuran) - tablet, powder for injection
colchicine (Colcrys) - tablet

Rheumatoid Arthritis Treatment

Etanercept binds and inactivates tumor necrosis factors (TNF), thereby preventing synovial inflammation.

etanercept (Enbrel) - injectable solution, prefilled syringe

PDE-5 Inhibitors

A phosphodiesterase type 5 inhibitor (PDE-5 inhibitor) is a drug used to block the degradative action of phosphodiesterase type 5 on cyclic GMP in the smooth muscle cells lining the blood vessels supplying the corpus cavernosum of the penis. These drugs are used in the treatment of erectile dysfunction, and were the first effective oral treatment available for the condition. Because PDE-5 is also present in the arterial wall smooth muscle within the lungs, PDE-5 inhibitors have also been used in the treatment of pulmonary hypertension, a disease in which blood vessels in the lungs become overloaded with fluid, usually as a result of failure of the left ventricle of the heart.

sildenafil (Viagra, Revatio) - tablet, injectable solution
tadalafil (Cialis, Adcirca) - tablet
vardenafil (Levitra, Staxyn ODT) - tablet, orally disintegrating tablet

Antitussives and Expectorants

A cough medicine is a medication used to treat or ameliorate coughing and related conditions. For dry coughs, treatment with cough suppressants, antitussives (benzonatate, codeine, dextromethorphan), may be attempted to suppress the body's urge to cough. However, in productive coughs (coughs that produce phlegm) treatment is instead attempted with expectorants (guaifenesin) in an attempt to loosen mucus from the respiratory tract.

benzonatate (Tessalon, Tessalon Perles) - capsule, gelcaps
codeine & guaifenesin (Cheratussin AC, Guiatuss AC) CV - oral liquid
dextromethorphan (Benylin DM) - oral suspension, gelcaps, capsule, syrup, lozenge
guaifenesin (Mucinex) - caplet, extended-release tablet, oral syrup
guaifenesin & phenylephrine (Sudafed PE Non-Drying Sinus Caplets) - caplet

Prostaglandins

Latanoprost is a prostaglandin F2-alpha analog, which causes an increase in the outflow of aqueous humor. This reduces intraocular pressure.

latanoprost (Xalatan) - ophthalmic solution

Estrogens and Progesterones

Estrogen by itself (including conjugated estrogens and estradiol) are used for menopausal vasomotor symptoms, atrophic vaginitis, Kraurosis vulvae, female hypogonadism, osteoporosis prevention, prostate cancer, abnormal uterine bleeding, female castration, primary ovarian failure, and breast cancer.

Progesterone only drugs (levonorgestrel, medroxyprogesterone, and norethindrone acetate) are used for amenorrhea, contraception, endometriosis, immune thrombocytopenic purpura, and paraphilia.

Estrogen and progesterone combinations (desogestrel & ethinyl estradiol, drospirenone & ethinyl estradiol, etonogestrel & ethynil estradiol, norelgestromin & ethinyl estradiol, norethindrone acetate & ethinyl estradiol, and norgestimate & ethinyl estradiol) are used for contraception, premenstrual dysphoric disorder, dysmenorrhea, endometriosis, and polycystic ovarian syndrome. Many of these combinations are also used for acne.

conjugated estrogens (Premarin) - tablet, powder for injection
desogestrel & ethinyl estradiol (Desogen, Apri, Micrette) - tablet
drospirenone & ethinyl estradiol (Yasmin, Yaz, Ocella) - tablet
estradiol (Estrace, Vivelle-Dot, Estraderm, Estring, Femring) - gel, injectable solution, tablet, transdermal patch, topical emulsion, vaginal cream, vaginal ring, vaginal tablet
etonogestrel & ethynil estradiol (NuvaRing) - intrauterine device
levonorgestrel (Mirena, Plan B) - intrauterine device, tablet
medroxyprogesterone (DepoProvera, Provera) - tablet, injectable suspension, prefilled syringes
norelgestromin & ethinyl estradiol (Ortho Evra, Evra Transdermal System) - transdermal patch
norethindrone acetate (Aygestin) - tablet
norethindrone acetate & ethinyl estradiol (Loestrin, Microgestin, Junel) - tablet
norgestimate & ethinyl estradiol (Ortho Cyclen, Ortho Tri-Cyclen, Ortho Tri-Cyclen Lo, Trinessa) - tablet

Quick review question 40

Which of the following medications affect the hemopoietic stem cells in bone marrow causing an increase in white blood cells?

   a. Aranesp
   b. Epogen
   c. Neupogen
   d. Neumega

Quick review question 41

Which of the following medications is considered an expectorant?

   a. benzonatate
   b. codeine

c. dextromethorphan
d. guaifenesin

Quick review question 42

Conjugated estrogens are used for which of the following?

a. menopausal vasomotor symptoms
b. contraception
c. acne
d. all of the above

Dietary supplements

Medications, biological agents, and injectables are heavily regulated by the Food and Drug Administration (FDA), whereas dietary supplements intended for oral consumption are only regulated as food by the FDA.

A dietary supplement is intended to supply nutrients (vitamins, minerals/electrolytes, and amino acids) that are missing or not consumed in sufficient quantity in a person's diet. This category may also include herbal supplements. Dietary supplements often make health related structure function claims.

Micronutrients

A micronutrient is a substance needed only in small amounts for normal body function (vitamins and minerals).

Vitamins

Vitamins can be defined as any of a group of substances that are essential, in small quantities, for the normal functioning of metabolism in the body. They cannot usually be synthesized in the body, but they occur naturally in certain foods. Insufficient supply of any particular vitamin results in a deficiency disease, hence the common need for vitamin supplements. Vitamins A, D, E, and K are fat soluble and therefore can be stored in adipose tissue. Other vitamins are considered water soluble.

multivitamin infusion (M.V.I.-12, Infuvite) - injectable solution -

Multivitamin infusions are commonly used for parenteral nutrition or for treatment of a nutritional deficiency.

prenatal vitamins (PreNexa, Nexa, NesTabs) - tablet, capsule - Prenatal vitamins are intended for use by women while trying to conceive and throughout their pregnancy to ensure adequate vitamin intake.

vitamin A, beta carotene (Retinol A, Aquasol A) - capsule, injectable solution, tablet - Vitamin A affects growth, development, and maintenance of epithelial tissue. Vitamin A deficiency causes night blindness, dry eyes, susceptibility to infections, and follicular hyperkeratosis.

vitamin B1, thiamine - tablet, injectable solution - Vitamin B1 affects carbohydrate metabolism and treats/prevents beriberi.

vitamin B3, niacin, nicotinic acid (Niacor, Niaspan) - tablet, extended-release tablet - Vitamin B3 effects lipid metabolism making it a treatment for hyperlipidemia.

vitamin B9, folic acid (Folvite) - cream, ointment, tablet, injectable solution. Vitamin B9 is necessary for red blood cell formation and prevents neural tube defects.

vitamin B12, cyanocobalamin (Cobex) - tablet, injectable solution, nasal spray - Vitamin B12 affects red blood cell production and can treat pernicious anemia.

vitamin C, ascorbic acid - tablet, extended-release tablet, capsule, extended-release capsule, oral solution, injectable solution - Vitamin C is required for collagen biosynthesis. Vitamin C deficiency can cause scurvy.

vitamin D2, ergocalciferol (Drisdol, Calciferol) - tablet, capsule, oral solution - Vitamin D affects maintenance of calcium & phosphorus homeostasis, cellular differentiation, gene regulation, and membrane integrity. Vitamin D deficiency leads to rickets in kids, osteomalacia in adults, and low calcium levels.

vitamin D3, cholecalciferol - tablet, capsule, oral solution - Vitamin D affects maintenance of calcium & phosphorus homeostasis, cellular differentiation, gene regulation, and membrane integrity. Vitamin D deficiency leads to rickets in kids, osteomalacia in adults, and low calcium levels.

vitamin E, tocopherol (Aquasol E) - tablet, capsule, oral solution - Vitamin E functions as a lipid antioxidant, protects membrane phospholipids, intracellular antioxidant, and inhibits platelet aggregation. Vitamin E has been used to treat skin conditions, leg cramps, sexual dysfunction, heart disease, aging, PMS, and to increase athletic performance.

vitamin K1, phytonadione (Mephyton, AquaMephyton) - tablet, injection emulsion - Vitamin K1 is an essential lipid-soluble vitamin that plays a vital role in the production of coagulation proteins. Vitamin K1 can be used as an antidote for warfarin.

Electrolytes

Electrolytes when dissolved in solution break apart into ions (anions and cations). The movement of these ions are needed by cells to regulate the flow of water molecules across cell membranes.

calcium with various salts (OS-Cal, Tums, PhosLo, Citracal) - tablet, effervescent tablet, chewable tablet, injectable solution, injectable suspension - Calcium is used to prevent osteoporosis and kidney stones and to treat hypocalcemia, hyperkalemia, hypermagnesemia, calcium channel blocker overdose, and can also be used as an antacid.

iron, ferrous with various salts (Slow Fe, Feosol, Ferrlecit, INFeD) - tablet, extended-release tablet, oral solution, oral liquid drops, oral suspension drops, injectable solution - Iron is necessary for red blood cell formation. Iron is used to treat iron deficient anemia.

magnesium with various salts (Mag-Ox 400, Uro-Mag, Magonate) - tablet, capsule, powder for oral solution, oral solution, injectable solution, infusion solution - Magnesium is one of the major intracellular cations. For normal neuromuscular activity, humans need normal concentration of extracellular calcium and magnesium. Intracellular magnesium is an important cofactor for various enzymes, transporters, and nucleic acids that are essential for normal cellular function, replication, and energy metabolism.

potassium with various salts (Klor-Con, KDur, Slow K, K Phos) - tablet, extended-release tablet, extended-release capsule, effervescent tablet, injectable solution, premix infusion - Potassium is used to treat or prevent hypokalemia.

zinc - tablet, capsule, gum, lozenge, ointment, paste, injectable solution - Zinc is an essential nutritional requirement and serves as a cofactor for more than 70 different enzymes, including carbonic anhydrase, alkaline phosphatase, lactic dehydrogenase, and both RNA and DNA polymerase. Zinc facilitates wound healing, helps maintain normal growth rates, normal skin hydration, and the senses of taste and smell. Zinc is also used to topically treat diaper rash and various topical irritations. Some oral types are used to ameliorate cold symptoms.

multiple trace elements (Multitrace, M.T.E.) - injectable solution - Multiple trace elements are used to provide various trace elements in a parenteral infusion.

Herbals

Herbal supplements are often used for an array of purposes. A common challenge to the use of herbal supplements is the lack of scientific studies for their use; however, there is a plethora of anecdotal evidence. The following is a brief list of the most commonly used herbal supplements.

chondroitin - Chondroitin is often used for osteoarthritis, particularly of the knee.

coenzyme Q10, CoQ10 - Coenzyme Q10 is used to treat angina, chronic fatigue syndrome, CHF, diabetes mellitus, doxorubicin-induced cardiotoxicity (prevention), HIV/AIDS immunostimulant, hypertension, mitochondrial cytopathies, muscular dystrophies, myopathy (statin-induced).

cranberry - Cranberry is used for UTI prevention and as a urinary deodorizer for incontinent patients.

creatine - Creatine is used for myotrophic lateral sclerosis, CHF, exercise performance enhancement, gyrate atrophy, McArdle disease, mitochondrial cytopathies, muscle mass builder, muscular dystrophies, neuromuscular disease, rheumatoid arthritis, and Parkinson's disease.

fish oil - Fish oil is used for coronary heart disease, hyperlipidemia, HTN, hypertriglyceridemia, Raynaud's syndrome, rheumatoid arthritis, and stroke prevention.

glucosamine - Glucosamine is used for relief of symptoms of osteoarthritis , and temporomandibular joint arthritis.

marijuana - Marijuana is used to treat decrease intraocular pressure, analgesia, antiemetic effects, and as an appetite stimulant.

phytoestrogens - Phytoestrogens are used to treat menopausal vasomotor symptoms, osteoporosis, decrease risk of breast cancer, and cardiovascular disease.

St. John's wort - St. John's wort is used to treat depression (mild-moderate), psychosomatic disorder, obsessive-compulsive disorder, anxiety, premenstrual syndrome, burning mouth syndrome, and neuropathy.

wild yam - Wild yam is used for estrogen replacement therapy, painful menstruation, libido, breast enlargement, diverticulosis, dysmenorrhea, estrogen replacement, gallbladder colic, libido enhancement, osteoporosis, postmenopausal vaginal dryness, premenstrual syndrome, and rheumatoid arthritis.

Quick review question 43

This vitamin has antihemorrhagic properties that make it useful as an antidote to warfarin?

  a. vitamin B9
  b. vitamin B12
  c. vitamin E
  d. vitamin K1

Vaccines

A vaccine is a biological preparation that improves immunity to a particular disease. A vaccine typically contains an agent that resembles a disease-causing microorganism, and is often made from weakened or killed forms of the microbe, its toxins, or one of its surface proteins. The agent stimulates the body's immune system to recognize the agent as foreign, destroy it, and "remember" it, so that the immune system can more easily recognize and destroy any of these microorganisms that it later encounters. Vaccines do not guarantee complete protection from a disease. The efficacy or performance of the vaccine is dependent on a number of factors, including the disease itself (for some diseases, vaccination performs better than for other diseases); the strain of vaccine (some vaccinations are for different strains of the disease); whether one kept to the timetable for the vaccinations; individuals who are "non-responders" (they do not generate antibodies even after being vaccinated correctly), or other factors such as ethnicity, age, or genetic predisposition. When a vaccinated individual does develop the disease vaccinated against, the disease is likely to be milder than without vaccination.

A common allergy concern for the influenza virus vaccine is that it is produced in chicken embryos and therefore contains some egg proteins. If a patient has an egg allergy, it may be advisable to avoid the influenza vaccine.

diphtheria & tetanus toxoids (Decavac, Tenivac, Td, DT) - IM injection
haemophilus influenza type b vaccine (ActHIB, Hiberix) - IM injection
hepatitis a vaccine inactivated (Havrix, Vaqta) - IM injection
hepatitis b vaccine (Engerix B, Recombivax HB) - IM injection
human papillomavirus vaccine (Gardasil) - IM injection
influenza virus vaccine (Fluarix, Fluzone, Afluria, FluMist) - IM injection, ID injection, Nasal Spray
measles mumps and rubella vaccine (M-M-R-II) - SC injection
meningococcal A C Y and W-135 diphtheria conjugate vaccine (Menactra) - IM injection
pneumococcal vaccine (Pneumovax 23, Prevnar 13) - SC/IM injection
poliovirus vaccine inactivated (IPOL) - SC/IM injection
rabies vaccine (HDCV, Imovax) - IM injection
rotavirus oral vaccine (Rotarix) - oral solution
tetanus & reduced diphtheria toxoids/ acellular pertussis vaccine (Adacel, Boostrix, Tdap) - IM injection
varicella virus vaccine live (Varivax) - SC injection
zoster vaccine live (Zostavax) - SC injection

Quick review question 44

If a patient has an allergy to eggs, which of the following vaccines may be a concern?

a. ActHIB
b. Gardasil
c. Fluarix
d. Pneumovax 23

Answers to quick review questions

1. C - Non-proprietary name is another name for a generic product.
2. B - If the products are not considered therapeutic

equivalents, they will be given a TE Code of "B".

3. D - Tetracyclines easily bind with magnesium, aluminum, iron, and calcium, which reduces its ability to be completely absorbed by the body. Therefore, none of the options were appropriate.

4. A - Keflex (cephalexin) is a cephalosporin antibiotic. Cephalosporins have a similar chemical structure to penicillin creating a potential for cross sensitivity.

5. C - Zosyn is the brand name for the combination drug piperacillin & tazobactam.

6. C- Zovirax (acyclovir) is classified as an antiviral medication.

7. B - Sulfonamides (such as sulfamethoxazole & trimethoprim) have the potential to crystallize in the kidneys due to their low solubility. This is a very painful experience, so patients are recommended to take these medication with large amounts of water.

8. D - Long term use of glucocorticoids can have the following negative effects: thinning of skin, decreases wound healing, stunting pediatric growth, moon-face, obesity, and diabetes mellitus.

9. B - Epinephrine, due to its heavy stimulation of the beta-2 adrenergic receptor, is used to treat sever asthma attacks (usually via a nebulizer).

10. A - Toprol XL is the brand name for the extended release form of metoprolol and is a beta-adrenergic blocking drug, which is a class of drugs sometimes simply referred to as beta blockers.

11. A - Neostigmine is often used to reverse the effects on nondepolarizing neuromuscular blocking agents. Various peripherally acting skeletal muscle relaxants like cisatracurium, pancuronium, rocuronium, and vecuronium achieve muscle relaxation by being nondepolarizing neuromuscular blocking agents.

12. D - All patients using skeletal muscle relaxants should avoid additional items that will depress the CNS or impair neuromuscular function, such as alcohol, sedatives, and tranquilizers.

13. B - Cocaine is the only local anesthetic that causes vasoconstriction.

14. C - Antidepressants carry a black box warning that in

short-term studies, antidepressants increased the risk of suicidal thinking and behavior in children, adolescents, and young adults.

15. D - Benzodiazepines are typically used as hypnotics and not for treating depression.

16. A - Olanzapine is an antipsychotic agent that can be used for treating schizophrenia. Clonazepam is a barbiturate commonly used as a hypnotic; zaleplon is commonly used as a sleep aid; and varenicline is used for smoking cessation.

17. D - Xanax (alprazolam), Luminal (phenobarbital), and Ambien (zolpidem) are all schedule IV controlled substances.

18. D - All of the medications in this category can be used to treat seizures.

19. C - Parkinson's disease is caused by the loss of dopamine receptors.

20. B - Buprenorphine & naloxone can be used as either a pain reliever or to treat opioid addiction. Physicians prescribing this medication to treat addiction are required to use a special DEA number that starts with an 'X'.

21. C - Ketorolac is available as both a tablet and an injectable solution.

22. A - Serotonin 5-HT receptor agonists are used to treat migraines and cluster headaches.

23. C - Aldactone (spironolactone) is considered a potassium sparing diuretic. Loop diuretics, like Lasix (furosemide), and thiazide diuretics, like hydrochlorothiazide, will cause patients to lose potassium. Tikosyn (dofetilide) is not a diuretic; it is an antiarrhythmic.

24. A - The generic name for Avapro is irbesartan. The other ARBs on the list pair up with brand names as follows: Cozaar (losartan), Micardis (telmisartan), and Diovan (valsartan).

25. B - Cardizem (diltiazem) is a calcium channel blocker. Capoten (captopril) is an angiotensin-converting enzyme inhibitor, Coreg (carvedilol) is a beta blocker, and Dilatrate-SR (isosorbide dinitrate) is a nitrate.

26. D - Coumadin (warfarin) is available as a tablet. Fragmin (dalteparin), Lovenox (enoxaparin), and heparin are only

available as injectables.

27. D - Lipitor (atorvastatin) decreases total levels of cholesterol in the blood, particularly LDL-cholesterol and triglycerides. It also increases HDL cholesterol.

28. B - Diphenhydramine can cross the blood brain barrier causing drowsiness and may therefore be used to treat insomnia.

29. C - Spiriva (tiotropium) requires a special HandiHaler to deliver the capsules as an inhalation.

30. A - Scopolamine (Transderm Scop) is available as tablet, transdermal patch, ophthalmic solution, and injectable solution.

31. D - Pepcid (famotidine) is an H2-receptor antagonist. Protonix (pantoprazole), Prilosec (omeprazole), and Prevacid (lansoprazole) are all considered proton pump inhibitors.

32. C - Paget's disease affects bone mass and development.

33. B - Lansoprazole is a proton pump inhibitor affecting acid production in the GI tract. Levothyroxine, thyroid, and liothyronine are all used to treat hypothyroidism.

34. A - Byetta (exenatide) is used to treat type 2 diabetes, but not type 1 diabetes. Insulins, including Apidra (insulin glulisine), Lantus (insulin glargine), and Humulin R (insulin regular human), can be used in the treatment of both type 1 and type 2 diabetes.

35. C - Amaryl (glimiperide) is a sulfonylurea and are closely related to sulfonamides causing cross sensitivity.

36. C - A type 1 diabetic requires both basal and prandial insulin.

37. D - Metformin decreases hepatic glucose production, decreases GI glucose absorption, and increases target cell insulin sensitivity.

38. D - Common side effect of chemotherapy often results in anemias, nausea, vomiting, diarrhea, hair loss, photosensitivity, and thinner skin.

39. A - Monoclonal antibodies are considered targeted therapy as they only work against specific cells.

40. C - Neupogen (filgrastim) will cause an increase in white blood cells. Aranesp (darbepoeitin) and Epogen (erythropoeitin) will cause an increase in red blood cells. Neumega (oprelvekin) will cause an increase in platelets.

41. D - Guaifenesin is an expectorant; whereas benzonatate, codeine, and dextromethorphan are considered antitussives.
42. A - Conjugated estrogens are used for treating menopausal vasomotor symptoms.
43. D - Vitamin K1, phytonadione (Mephyton, AquaMephyton) plays a vital role in the production of coagulation proteins and can be used as an antidote for warfarin.
44. C - Fluarix is an influenza virus vaccine. A common allergy concern for the influenza virus vaccine is that it is produced in chicken embryos and therefore contains some egg proteins.

CHAPTER 2 Pharmacy Law and Regulation

Key concepts

This chapter will cover the following knowledge areas to prepare you for the *Pharmacy Technician Certification Exam*:

- Storage, handling, and disposal of hazardous substances and wastes (e.g., MSDS)
- Hazardous substances exposure, prevention and treatment (e.g., eyewash, spill kit, MSDS)
- Controlled substance transfer regulations (DEA)
- Controlled substance documentation requirements for receiving, ordering, returning, loss/theft, destruction (DEA)
- Formula to verify the validity of a prescriber's DEA number (DEA)
- Record keeping, documentation, and record retention (e.g., length of time prescriptions are maintained on file)
- Restricted drug programs and related prescription-processing requirements (e.g., thalidomide, isotretinoin, clozapine)
- Professional standards related to data integrity, security, and confidentiality (e.g., HIPAA, backing up and archiving)
- Requirement for consultation (e.g., OBRA'90)
- FDA's recall classification
- Infection control standards (e.g., laminar air flow, clean room, hand washing, cleaning counting trays, countertop, and equipment) (OSHA, USP 795 and 797)
- Record keeping for repackaged and recalled products and supplies (TJC, BOP)
- Professional standards regarding the roles and responsibilities of pharmacists, pharmacy technicians, and other pharmacy employees (TJC, BOP)
- Reconciliation between state and federal laws and

regulations
- Facility, equipment, and supply requirements (e.g., space requirements, prescription file storage, cleanliness, reference materials) (TJC, USP, BOP)

Terminology

To get started with, there is some terminology that should be defined.

hazardous drugs - Hazardous drugs are drugs that are known to cause genotoxicity, which is the ability to cause a change or mutation in genetic material; carcinogenicity, the ability to cause cancer in animal models, humans or both; teratogenicity, which is the ability to cause defects on fetal development or fetal malformation; and lastly, hazardous drugs are known to have the potential to cause fertility impairment, which is a major concern for most clinicians. These drugs can be classified as antineoplastics, cytotoxic agents, biologic agents, antiviral agents, and immunosuppressive agents.

Occupational Safety and Health Administration (OSHA) - OSHA is a government agency within the United States Department of Labor responsible for maintaining safe and healthy work environments.

Material Safety Data Sheet (MSDS) - OSHA-required notices on hazardous substances which provide hazard, handling, clean-up, and first aid information.

personal protective equipment (PPE) - Personal protective equipment is worn by an individual to provide both protection to the wearer from the environment or specific items they are manipulating, and to prevent exposing the environment or the items being manipulated directly to the wearer of the PPE.

Drug Enforcement Administration (DEA) - The Drug Enforcement Administration (DEA) is a United States Department of Justice law enforcement agency, a federal police service tasked with enforcing the Controlled Substances Act of 1970.

controlled substances - Controlled substances are medications with restrictions due to abuse potential. There are 5 schedules of controlled substances with various prescribing guidelines based on abuse potential, counter balanced by potential medicinal benefit as determined by the Drug Enforcement Administration and individual state legislative branches.

Risk Evaluation and Mitigation Strategy (REMS) - The Food and Drug Administration may require risk evaluation and mitigation strategies (REMS) on medications, if necessary, to minimize the risks associated with some drugs. REMS may require a medication guide, a communications plan, elements to assure safe use, an implementation plan, and a timetable for submission of assessments.

United States Pharmacopeia (USP) - The United States Pharmacopeia (USP) is the official pharmacopeia of the United States, published dually with the National Formulary as the USP-NF. Prescriptions and over-the-counter medicines and other healthcare products sold in the United States are required to follow the standards in the USP-NF. The USP also sets standards for food ingredients and dietary supplements.

compounded sterile preparations (CSP) - Compounded sterile preparations are admixtures that need to be assembled under aseptic conditions to prevent contamination.

Quick review question 1

Which of the following could be caused by hazardous drugs?

   a. they could cause cancer
   b. they could damage or mutate genetic material
   c. they could harm a developing fetus and/or cause fertility impairment
   d. all of the above

Quick review question 2

Which organization requires pharmacies to maintain MSDS?

   a. OSHA
   b. FDA
   c. DEA
   d. USP

Safety requirements

There is a vast array of information to be aware of with safety requirements posed by various organizations. Of particular concern is the safe storage, handling, managing accidental exposure, and disposal of hazardous materials. Hazardous drugs are drugs that are known to cause genotoxicity, which is the ability to cause a change or mutation in genetic material; carcinogenicity, which is the ability to cause cancer in animal models, humans or both; teratogenicity, which is the ability to cause defects on fetal development or fetal malformation; and lastly, hazardous drugs are known to have the potential to cause fertility impairment, which is a major concern for most clinicians. These drugs can be classified as antineoplastics, cytotoxic agents, biologic agents, antiviral agents, and immunosuppressive agents. This is why safe handling of hazardous drugs is crucial.

Product storage

The safety requirements include everything from the proper inventory rotation to avoid dispensing expired products, to material safety data sheets to provide the necessary information for safe clean up after accidental spills, to appropriate handling of oncology materials, and proper storage of chemicals and flammable items.

Proper rotation of inventory and periodic checking of expirations help to reduce the potential for dispensing expired medications. It also maximizes the utilization of inventory before medications become outdated. When looking at expirations on medication vials, it is important to note that if a medication only mentions the month and year, but not the day, then you are to treat it as expiring at the end of the month. As an example, if a medication is marked as expiring on 02/2020, then you would treat it as expiring on February 29, 2020.

The Occupational Safety and Health Administration (OSHA) requires all work places, including pharmacies, to carry material safety data sheets (MSDS) for all hazardous substances that are stored on the premises. This includes oncology drugs and volatile chemicals along with other hazardous chemicals. The MSDS provide handling, clean-up, and first-aid information.

Segregating inventory by drug categories helps to prevent potentially harmful errors. The Joint Commission (TJC), formerly known as the Joint Commission on Accreditation of Healthcare Organizations (JCAHO), requires that internal and external medications must be stored separately. This reduces the potential that someone will dispense or administer an external product for internal use. The Joint Commission also has requirements for separate storage of oncology drugs and volatile or flammable substances.

Hazardous drugs (i.e., oncology drugs) should have a separate space on the shelves and be labeled in such a way that it will alert staff of the hazardous potential of these medications. Oncology drugs are often cytotoxic themselves and must be handled with extreme care. They should be received in a sealed protective outer bag that restricts dissemination of the drug if the container leaks or is broken. When potential exists for exposure to hazardous drugs, all personnel involved must wear appropriate personal protective equipment while following a hazardous materials cleanup procedure. All exposed materials must be properly disposed of in hazardous waste containers.

Volatile or flammable substances (including tax free alcohol) require careful storage. They must have a cool location that is properly ventilated. Their storage area must be designed to reduce fire and explosion potential.

Product handling

Whenever receiving and/or handling hazardous materials, individuals are recommended to use personal protective equipment (PPE). PPE typically consists of booties, hair covers, a respirator (typically a HEPA mask), face shield, coated gowns capable of providing at least 4 hours of protection if accidentally exposed, powder free impermeable gloves (either double gloved and/or chemotherapy gloves). The American Society of Healthsystem Pharmacists (ASHP) and the National Institute of Occupational Safety and Health (NIOSH) both recommend changing gowns every three hours during extended work with hazardous substances. It is also recommended that gloves be changed every 30 minutes.

Accidental exposure

Workers may be exposed to a hazardous drug at any point during its manufacture, transport, distribution, receipt, storage, preparation, and administration, as well as during waste handling, and equipment maintenance and repair. All workers involved in these activities have the potential for contact with uncontained drug.

In case of accidental exposure, you should consult the Material Safety Data Sheet (MSDS) which provide hazard, handling, clean-up, and first aid information. An eyewash is an appropriate device to utilize if hazardous or other unwanted materials come in contact with the eye. The individual exposed will probably be required to file an initial incident report (IIR) with their facility.

Facilities that utilize hazardous drugs are required to have spill kits, which may be useful in the containment and clean-up of hazardous materials. Spill kits typically contain PPE, waste containers, warning signs to help minimize traffic through the contaminated area, powders for solidifying liquids, disposable scoops and brushes, absorbent gauze or pads, and a common deactivating agent such as sodium hypochlorite.

Proper waste disposal

Properly labeled, leak-proof, and spill-proof containers of nonreactive plastic are required for areas where hazardous waste is generated, and can be further broken down into either yellow or black containers. Fully used vials, syringes, tubing, and bags of hazardous drug waste, along with PPE used while working with hazardous drugs, may be disposed in yellow, properly labeled containers. Also, any partially used or expired hazardous drugs that are not also considered to be RCRA (Resource Conservation and Recovery Act) regulated hazardous drugs should be placed in these yellow buckets. Trace contaminated items such as booties, gowns, and masks, even if they were not involved in spills, MUST be treated as hazardous waste (the yellow buckets are sufficient for this).

Partially used bags, vials, syringes, etc. of RCRA listed hazardous pharmaceutical waste MUST be placed in black RCRA approved containers. Hazardous waste must be properly manifested and transported by a federally permitted hazardous waste transporter to a federally permitted hazardous waste storage, treatment, or disposal facility.

A licensed contractor may be hired to manage the hazardous waste program.

Quick review question 3

If a medication is listed as expiring 10/2020, on what day does it expire?

a. on the last day of the previous month
b. at the beginning of the month
c. on the last day of the month
d. a specific day needs to be stated for a manufacturer's expiration to be valid

Quick review question 4

Why does TJC require internal and external products to be stored separately?

a. because internal products typically come in unit of use, whereas external products tend to come in bulk packages
b. it reduces the potential that someone will use an external product internally
c. to make it easier to find a patient's medications
d. none of the above

Quick review question 5

Whenever receiving and/or handling hazardous materials you should use:

a. PPE
b. a positive pressure buffer room
c. USP
d. all of the above

Quick review question 6

When may workers be exposed to hazardous drugs?

a. during its preparation
b. during its distribution

c. during waste handling
d. all of the above

Quick review question 7

In a healthcare facility, where should RCRA listed waste be disposed of?

a. in a red sharps container
b. in a yellow hazardous waste container
c. in a black RCRA approved container
d. in the general trash as long as it has been deidentified

Controlled substances regulations

The mission of the Drug Enforcement Administration (DEA) is to enforce the controlled substances laws and regulations of the United States and bring to the criminal and civil justice system of the United States, or any other competent jurisdiction, those organizations and principal members of organizations involved in the growing, manufacture, or distribution of controlled substances appearing in or destined for illicit traffic in the United States; and to recommend and support non-enforcement programs aimed at reducing the availability of illicit controlled substances on the domestic and international markets.

In carrying out its mission as the agency responsible for enforcing the controlled substances laws and regulations of the United States it directly impacts scheduling abuse potential, tracking, receiving, ordering, returning, loss/theft, and destruction of controlled substances.

Medication schedules

Controlled substances are medications with further restrictions due to abuse potential. There are 5 schedules of controlled substances with various prescribing guidelines based on abuse potential counter balanced by potential medicinal benefit as determined by the Drug Enforcement Administration and individual state legislative branches. The DEA is provided with this authority by the Controlled

Substances Act. Below is a brief explanation of the schedules along with example medications.

Schedule I (CI) medications have unaccepted medical use and the highest potential for abuse. These are not available by a prescription. Some examples of CI medications include LSD, heroin, and Quaaludes (methaqualone). CI medications are not available via a prescription.

Schedule II (CII) medications have a high potential for abuse or misuse but sufficient medicinal use to justify availability as a prescription. Some examples of CII medications include oxycodone, morphine, and amphetamines. CII medications may be written for a maximum 90 day supply without refills.

Schedule III (CIII) medications have a potential risk for abuse, misuse, and dependence. Some examples of CIII medications include Vicodin (hydrocodone bitartrate and acetaminophen), and codeine containing products in a solid dosage form (tablet, capsule, etc.). CIII – CV medications may be written for up to a 6 month supply.

Schedule IV (CIV) medications have a low potential for abuse and limited risk of dependence. Some examples of CIV medications include phenobarbital, benzodiazepines, and other sedatives and hypnotics.

Schedule V (CV) medications have a low potential for abuse or misuse. Some examples of CV medications include cough medicines that contain a limited amount of codeine, and antidiarrheal medications that contain a limited amount of an opiate such as Lomotil (diphenoxylate and atropine).

Many problems associated with drug abuse are the result of legitimately-manufactured controlled substances being diverted from their lawful purpose into the illicit drug traffic. Many of the narcotics, depressants and stimulants manufactured for legitimate medical use are subject to abuse, and have therefore been brought under legal control. The goal of controls is to ensure that these "controlled substances" are readily available for medical use, while preventing their distribution for illicit sale and abuse.

Diversion control system

Many problems associated with drug abuse are the result of legitimately-manufactured controlled substances being diverted from their lawful purpose into the illicit drug traffic. Many of the narcotics, depressants and stimulants manufactured for legitimate medical use are subject to abuse, and have therefore been brought under legal control. The goal of controls is to ensure that these "controlled substances" are readily available for medical use, while preventing their distribution for illicit sale and abuse.

Under federal law, all businesses which manufacture or distribute controlled drugs, all health professionals entitled to dispense, administer or prescribe them, and all pharmacies entitled to fill prescriptions must register with the DEA. Authorized registrants receive a "DEA number". Registrants must comply with a series of regulatory requirements relating to drug security, records accountability, and adherence to standards.

A DEA number is a series of numbers assigned to a health care provider (such as a dentist, physician, nurse practitioner, or physician assistant) allowing them to write prescriptions for controlled substances. Legally the DEA number is solely to be used for tracking controlled substances. The DEA number, however, is often used by the industry as a general "prescriber" number that is a unique identifier for anyone who can prescribe medication.

A valid DEA number consists of 2 letters and 7 digits. The first letter is always an A (deprecated), B (most common), or F (new) for a dispenser. The second letter is typically the initial of the registrant's last name. The seventh digit is a "checksum" that is calculated by adding together the first, third and fifth digits, then adding together the second, fourth and sixth digits, and multiply that sum by 2. Add both sums, and the last digit (the ones value) of this last sum is used as the seventh digit in the DEA number.

Purchasing

Controlled substances require special consideration when it comes to purchasing.

Schedule III – V drugs may be ordered by a pharmacy or other appropriate dispensary on a general order from a wholesaler and you should check the delivery in against the original order.

Schedule II drugs have much more stringent requirements. A pharmacy must register with the Drug Enforcement Administration (DEA) to purchase schedule II medications. The purchase of schedule II controlled substances must be authorized by a pharmacist and executed on either a triplicate DEA 222 order form or an electronic 222 form through a controlled substances ordering system (CSOS)

The printed DEA form 222 is a triplicate form. The pharmacy retains the third sheet while sending the first and second pages to the wholesaler. The wholesaler is responsible for sending the second page to the DEA while retaining the first page for its own records. When the Schedule II medications arrive in the pharmacy, they should be checked in against the DEA form.

Sample DEA Form 222

See Reverse of PURCHASER's Copy for Instructions		No order form may be issued for Schedule I and II substance unless completed application form has been received. (21 CRF 1305.04)		OMB APPROVAL NO. 1117-0010
TO:		STREET ADDRESS		

CITY and STATE		DATE	TO BE FILLED IN BY SUPPLIER	
			SUPPLIER DEA REGISTRATION No.	

	No. of Packages	Size of Package	Name of Item	National Drug Code	Packaging Shipped	Date Shipped
	TO BE FILLED IN BY PURCHASER					
1						
2						
3						
4						
5						
6						
7						
8						
9						
10						

LAST LINE COMPLETED ◀ *(MUST BE 10 OR LESS)* SIGNATURE OF PURCHASER OR ATTORNEY OR AGENT

Date Issued	DEA Registration No.	Name and Address of Registrant
Schedules .		
Registered as a	No. of this Order Form	

DEA Form 222
(Oct 1992)

US OFFICIAL ORDER FORMS - SCHEDULES I & II
DRUG ENFORCEMENT ADMINISTRATION
SUPPLIER'S Copy 1

The following is an image explaining how ordering schedule II medications work with a controlled substances ordering system (CSOS)

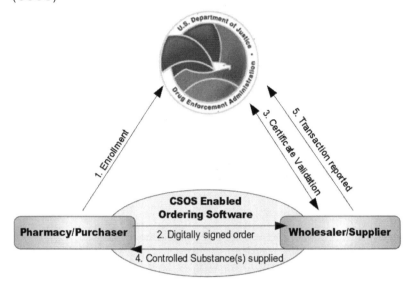

1. An individual enrolls with the DEA and, once approved, is issued a personal CSOS Certificate.
2. The purchaser creates an electronic 222 order using an approved ordering software. The order is digitally signed using the purchaser's personal CSOS Certificate and then transmitted to the suppliers. The paper 222 is not required for electronic ordering.
3. The supplier receives the purchase order and verifies that the purchaser's certificate is valid with the DEA. Additionally, the supplier validates the electronic order information just like it would a paper order.
4. The supplier completes the order and ships to the purchaser. Any communications regarding the order are sent electronically.
5. The order is reported by the supplier to the DEA within two business days.

Receiving

When the pharmacy receives controlled substances, they should be carefully checked in against the purchase order, including product name, quantity, strength, and package size. Controlled substances are shipped in separate containers from the rest of the pharmacy order and should be checked in by a pharmacist, although pharmacy technicians may assist with this process under the direct supervision of a pharmacist. Schedule II medications need to be checked in against your DEA 222 form (whether the paper triplicate form, or the electronic form on your CSOS enabled software).

Schedule II medications may be stocked separately in a secure place or disbursed throughout the stock. Some states may place further restrictions on how they may be stored. Their stock must be continually monitored and documented.

You may store CIII – CV medications in one of two ways:

- In a secured vault or,
- Dispersed throughout the pharmacy stock. By dispersing the stock throughout, you effectively prevent someone from being able to obtain all your controlled substances, since they cannot do easy *"One stop shopping."*

Transferring of controlled substances

A pharmacy may hire an outside firm to inventory, package, and arrange for the transfer of its controlled substances to another pharmacy, the original supplier, or the original manufacturer. The pharmacy is responsible for the actual transfer of the controlled substances and for the accuracy of the inventory and records. The records involving the transfer of controlled substances must be kept readily available by the pharmacy for two years for inspection by the DEA.

To transfer schedule II substances, the receiving registrant must issue an official order form (DEA Form 222) or an electronic equivalent to the registrant transferring the drugs. The transfer of schedules III-V controlled substances must be documented in writing to show the drug name, dosage form, strength, quantity, and date transferred. The document must include the names, addresses, and DEA registration numbers of the parties involved in the transfer of the controlled substances.

Transferring a prescription for a controlled substance

The DEA allows the transfer of original prescription information for Schedule III, IV, and V controlled substances for the purpose of refill dispensing between pharmacies on a one-time basis. Pharmacies which electronically share a real-time, online database may transfer up to the maximum number of refills permitted by the law and authorized by the prescriber. In either type of transfer, specific information must be recorded by both the transferring and the receiving pharmacist.

The DEA does not allow for the transfer of Schedule II controlled substances, as they do not allow refills on these medications, and partial fills, as discussed above, have strict limitations.

Disposal

A pharmacy may transfer controlled substances to a DEA registered reverse distributor who handles the disposal of controlled substances. The pharmacy should contact the local DEA Diversion Field Office for an updated list of DEA registered reverse distributors. In no case should drugs be forwarded to the DEA unless the registrant has received prior approval from the DEA. The DEA procedures established for the disposal of controlled substances must not be construed as altering in any way the state laws or regulations for the disposal of controlled substances.

Record keeping requirements

Every pharmacy must maintain complete and accurate records on a current basis for each controlled substance purchased, received, distributed, dispensed, or otherwise disposed of. These records must be maintained for 2 years. It is also required that records and inventories of Schedule II and Schedule III, IV, and V drugs must be maintained separately from all other records or be in a form that is readily retrievable from other records. The "readily retrievable" requirement means that records kept by automatic data processing systems or other electronic means must be capable of being separated out from all other records in a reasonable time. In addition, some notation, such as a 'C' stamp, an asterisk, red line, or other visually identifiable mark, must distinguish controlled substances from other items.

Inventory requirements

A pharmacy is required by the DEA to take an inventory of controlled substances every 2 years (biennially). This inventory must be done on any date that is within 2 years of the previous inventory date. The inventory record must be maintained at the registered location in a readily retrievable manner for at least 2 years for copying and inspection by the Drug Enforcement Administration. An inventory record of all Schedule II controlled substances must be kept separate from those of other controlled substances. Submission of a copy of any inventory record to the DEA is not required unless requested.

When taking the inventory of Schedule II controlled substances, an actual physical count must be made. For the inventory of Schedule III, IV, and V controlled substances, an estimate count may be made. If the commercial container holds more than 1000 dosage units and has been opened, however, an actual physical count must be made.

Theft or loss of a controlled substance

In the event that controlled substances are stolen, or are found to be missing, a pharmacy should:

1. contact the local police and the DEA,
2. fill out and file DEA form 106.

The DEA form 106 is available as both a printed form, or may be filed electronically.

Reconciliation between state and federal regulations

Many states have additional regulations for controlled substances. It is a necessary to know the differences that exist in your state. Some examples include:

- Some states have made additional drugs scheduled medications such as Arkansas and Kentucky have made tramadol a controlled substance.
- Many states, such as Pennsylvania, treat schedule V prescriptions the same as schedule III and IV prescriptions with regards to refill limits and totally quantity that may be prescribed at a time.
- Various states, such as New York, limit schedule II prescriptions to a 30 day supply.
- Some states, such as Hawaii and Montana, currently do not allow e-prescribing of controlled substances.

Be aware of the differences with regards to controlled substances in the states you practice in, and remember that the expectation when state and federal law appear to be in conflict is that you will follow the stricter regulation.

Quick review question 8

DEA stands for what?

a. Drug Enforcement Agency
b. Drug Enforcement Administration
c. Drug Enrichment Agency
d. Drug Enrichment Administration

Quick review question 9

Controlled substance schedules are primarily based on what?

a. efficacy
b. abuse potential
c. how long their half-life is
d. how long they can be prescribed for

Quick review question 10

Controlled substances are broken into how many schedules?

a. one
b. three
c. five
d. seven

Quick review question 11

Which of the following medications is not a CIV?

a. Ambien
b. Xanax
c. Luminal
d. Mobic

Quick review question 12

Which of the following DEA numbers could be appropriate for Donald Ferguson, M.D.?

a. BF6428521
b. DF3456325
c. AF1111116
d. FF642852141

Quick review question 13

Oxycontin would most likely be ordered for the pharmacy inventory using which of the following options?

a. a standard warehouse order
b. DEA form 106
c. DEA form 222
d. A prescription with a valid DEA number

Quick review question 14

Restoril would most likely be ordered for the pharmacy inventory using which of the following options?

a. a standard warehouse order
b. DEA form 106
c. DEA form 222
d. A prescription with a valid DEA number

Quick review question 15

Under federal law, how long must the records involving the transfer of controlled substances be kept readily available for the DEA?

  a. 180 days
  b. 2 years
  c. 5 years
  d. indefinitely

Quick review question 16

Under federal regulations, how many times may a prescription for Tylenol No. 3 be transferred?

  a. none
  b. once
  c. five times, but only once per month
  d. as often as the patient wants

Quick review question 17

How often is a pharmacy required by the DEA to take an inventory of controlled substances?

  a. perpetually
  b. biannually
  c. annually
  d. biennially

Restricted drug programs

There are a number of medications with restrictions due to patient safety concerns. The FDA primarily manages these medications through their Risk Evaluation and Mitigation Strategy (REMS) initiative. The FDA may require manufacturers of drugs with safety concerns to submit a REMS program at the time a new drug is approved. These programs may contain any combination of 5 criteria (Medication Guide, Communication Plan, Elements to Assure Safe Use, Implementation System, and Timetable for Submission of Assessments). Restricted access programs are considered Elements to Assure Safe Use. A current list of medications with REMS programs can be found at

Some of the medications with the most significant restrictions
include thalidomide, clozapine, buprenorphine, and isotretinoin.
These programs are intended to make sure that the patients are
using the medications appropriately and to monitor them for
undesirable side effects.

Thalidomide

Thalidomide was first developed in Germany as a sleep aid. Shortly
after entering the European market, the manufacturer decided it was
also good for treating morning sickness. Unfortunately, the drug
caused severe and life-threatening birth defects in 40% of infants. In
1998, thalidomide was approved for limited distribution in the United
States for treating multiple myeloma. It is also used to treat
erythema nodosum leprosum.

Thalidomide is only available through the THALOMID REMS
program. Prescribers must be certified with the program; patients
must comply with the strict guidelines of the program; and
pharmacies must be certified with the program. Pharmacies must
only dispense to patients who are authorized to receive the drug and
comply with the requirements.

Clozapine

Clozapine is an atypical antipsychotic medication used in the
treatment of schizophrenia, and is also sometimes used off-label for
the treatment of bipolar disorder. Clozapine carries many warnings,
including warnings for agranulocytosis, CNS depression, leukopenia,
neutropenia, seizure disorder, bone marrow suppression, dementia,
hypotension, myocarditis, orthostatic hypotension (with or without
syncope), and seizures.

With these severe side effects in mind, each of the manufacturers
are required to enroll patients taking their medication into a national
registry where they will monitor a patient's white blood cell count
(WBC) and their absolute neutrophil count (ANC). If the numbers fall

below a particular level their therapy may need to be halted, and the national registry has a responsibility to report the information to the nonrechallengeable database. Then, if a physician attempts to place the patient on the medication again the various manufacturers will know they are not allowed as they always need to check new orders for a patient against the nonrechallengeable database. Once listed in the nonrechallengeable database, they may never use clozapine again.

Buprenorphine

The Drug Addiction Treatment Act of 2000 (DATA 2000) permits physicians who meet certain qualifications to treat opioid addiction with Schedule III, IV, and V narcotic medications that have been specifically approved by the Food and Drug Administration for that indication. Such medications may be prescribed and dispensed by waived physicians in treatment settings other than the traditional Opioid Treatment Program (methadone clinic) setting.

Since there is only one narcotic medication approved by the FDA for the treatment of opioid addiction within the Schedules given, DATA 2000 basically refers to the use of buprenorphine for the treatment of opioid addiction. Methadone is a Schedule II narcotic approved for the same purpose within the highly regulated methadone clinic setting.

Under the Act, physicians may apply for a waiver to prescribe buprenorphine for the treatment of opioid addiction or dependence. Requirements include a current State medical license, a valid DEA registration number, specialty or subspecialty certification in addiction from the American Board of Medical Specialties, American Society of Addiction Medicine, or American Osteopathic Association. Exceptions were also created for physicians who participated in the initial studies of buprenorphine and for State certification of addiction specialists. However, the Act is intended to bring the treatment of addiction back to the primary care provider. Thus most waivers are obtained after taking an 8 hour course from one of the five medical organizations designated in the Act and otherwise approved by the Secretary of the Department of Health and Human Services. When a physician qualifies for the waiver, he is given a second DEA number that begins with an 'X'. This new DEA number is only to be used when prescribing buprenorphine for the purpose of treating addictions. Once a physician obtains the waiver, he or she may treat up to 30 patients for narcotic addiction with buprenorphine. Recent changes to DATA 2000 have increased the patient limit to 100 for physicians who have had their waiver for a year or more and request the higher limit in writing.

Isotretinoin

Isotretinoin is indicated for severe nodular acne unresponsive to conventional therapy in patients 12 years of age and older. Isotretinoin is a teratogen and is highly likely to cause birth defects if taken by women during pregnancy, or even a short time before conception. A few of the more common birth defects this drug can cause are hearing and visual impairment, missing or malformed earlobes, facial dysmorphism, and mental retardation.

Due to the teratogenicity of this product, isotretinoin is only available through a restricted program under REMS called iPLEDGE. Prescribers, patients, pharmacies, and distributors must enroll in the iPLEDGE program (https://www.ipledgeprogram.com).

Quick review question 18

Through what program does the FDA primarily manage medications with significant safety concerns?

a. Electronic Orange Book
b. REMS
c. S.M.A.R.T.
d. S.T.E.P.S.

Quick review question 19

Why is thalidomide considered a restricted drug?

a. genotoxicity
b. carcinogenicity
c. teratogenicity
d. its usefulness as a sleep aid

Quick review question 20

After a patient is entered in the nonrechallengeable database, how long must they wait before restarting treatment with clozapine?

   a. if the patient immediately starts taking colony stimulating factors, they may stay on clozapine
   b. as soon as lab values demonstrate that their WBC and ANC have gone back up to acceptable levels
   c. after 6 months with consistently good WBC and ANC levels
   d. never

Quick review question 21

What is the name of the restricted program under REMS under which isotretinoin patients must enroll?

   a. DATA 2000
   b. iPLEDGE
   c. S.M.A.R.T.
   d. S.T.E.P.S.

Quick review question 22

If a physician is prescribing buprenorphine for the purpose of treating addiction, the DEA number they use should start with what?

   a. either A, B, or F
   b. 3, representing buprenorphine's Schedule
   c. a tilde '~'
   d. X

Omnibus Budget Reconciliation Act of 1990

While most federal laws provide the pharmacist with guidance on handling pharmaceuticals, the Omnibus Budget Reconciliation Act of 1990 (OBRA-90) placed expectations on the pharmacist in how to interact with the patient. While the primary goal of OBRA-90 was to save the federal government money by improving therapeutic outcomes, the method to achieve these savings was implemented by imposing on the pharmacist counseling obligations, prospective drug utilization review (ProDUR) requirements, and record-keeping mandates.

The OBRA-90 ProDUR language requires state Medicaid provider pharmacists to review Medicaid recipients' entire drug profile before filling their prescription(s). The ProDUR is intended to detect potential drug therapy problems. Computer programs can be used to assist the pharmacist in identifying potential problems. It is up to the pharmacist's professional judgment, however, as to what action to take, which could include contacting the prescriber. As part of the ProDUR, the following are areas for drug therapy problems that the pharmacist must screen:

- Therapeutic duplication
- Drug–disease contraindications
- Drug–drug interactions
- Incorrect drug dosage
- Incorrect duration of treatment
- Drug–allergy interactions
- Clinical abuse/misuse of medication

OBRA-90 also required states to establish standards governing patient counseling. In particular, pharmacists must offer to discuss the unique drug therapy regimen of each Medicaid recipient when filling prescriptions for them. Such discussions must include matters that are significant in the professional judgment of the pharmacist. The information that a pharmacist may discuss with a patient is found in the following list below.

- Name and description of the medication.
- Dosage form, dosage, route of administration, and duration of drug therapy.
- Special directions and precautions for preparation, administration, and use by the patient.
- Common severe side effects or adverse effects or interactions and therapeutic contraindications that may be encountered.
- Techniques for self-monitoring of drug therapy.
- Proper storage.
- Refill information.
- Action to be taken in the event of a missed dose.

Under OBRA-90, Medicaid pharmacy providers also must make reasonable efforts to obtain, record, and maintain certain information on Medicaid patients. This information, including pharmacist comments relevant to patient therapy, would be considered reasonable if an impartial observer could review the documentation and understand what has occurred in the past, including what the pharmacist told the patient, information discovered about the patient, and what the pharmacist thought of the patient's drug therapy. Information that would be included in documented information are listed below.

- Name, address, and telephone number.
- Age and gender.
- Disease state(s) (if significant)
- Known allergies and/or drug reactions.
- Comprehensive list of medications and relevant devices.
- Pharmacist's comments about the individual's drug therapy.

While OBRA-90 was geared to ensure that Medicaid patients receive specific pharmaceutical care, the overall result of the legislation provided that the same type of care be rendered to all patients, not just Medicaid patients. The individual states did not establish 2 standards of pharmaceutical care - one for Medicaid patients and another for non-Medicaid patients. The end result is that all patients are under the same professional care umbrella requiring ProDUR, counseling, and documentation.

Quick review question 23

How does OBRA-90 improve therapeutic outcomes?

a. it imposes pharmacist counseling obligations
b. it requires ProDURs
c. it has record keeping mandates
d. all of the above

Quick review question 24

When looking at ProDURs a pharmacist must screen for all of the following except:

a. the physicians expertise in the disease state the patient is being treated

b. therapeutic duplication
c. incorrect drug dosage
d. clinical abuse/misuse of medication

Quick review question 25

Out of the following, which would not be considered counseling under OBRA-90?

a. special directions and precautions with the medication
b. common severe side effects
c. more affordable homeopathic remedies
d. techniques for self-monitoring of drug therapy

Health Insurance Portability and Accountability Act

The Health Insurance Portability and Accountability Act of 1996 (HIPAA) is the most significant piece of federal legislation to affect pharmacy practice since OBRA-90.

The Privacy Rule component of HIPAA took effect on April 14, 2003, and was the first comprehensive federal regulation designed to safeguard the privacy of protected health information (PHI). Pharmacies that maintain patient information in electronic format or conduct financial and administrative transactions electronically, such as billing and fund transfers, must comply with HIPAA.

While HIPAA places stringent requirements on pharmacies to adopt policies and procedures relating to the protection of patient PHI, the law also gives important rights to patients. These rights include the right to access their information, the right to seek details of the disclosure of information, and the right to view the pharmacy's policies and procedures regarding confidential information.

The Health Insurance Portability and Accountability Act (HIPAA) imposes 5 key provisions upon pharmacies.

1. The first provision is the requirement that each pharmacy take reasonable steps to limit the use of, disclosure of, and the requests for PHI. PHI is defined as individually

identifiable health information transmitted or maintained in any form and via any medium. To be in compliance, a pharmacy must implement reasonable policies and procedures that limit how PHI is used, disclosed, and requested for certain purposes. The pharmacy also is obligated to post its entire notice of privacy practices at the facility in a clear and prominent location and on its Web site (if one exists).

2. The second component of HIPAA requires that individuals be informed of the privacy practices of the pharmacy and that the pharmacy develop and distribute a notice with a clear explanation of these rights and practices. This notice must be given to every individual no later than the date of the first service provided, which usually means the first prescription dispensed to the patient. The pharmacist also is obligated to make a good-faith effort to obtain the patient's written acknowledgment of the receipt of the notice.

3. Under the third component, pharmacies are required, as well, to select a compliance officer who will manage and ensure compliance with HIPAA.

4. As part of the fourth component of HIPAA, all employees working in the pharmacy environment in which PHI is maintained must receive training on the regulations within a reasonable time after being hired. This training necessarily includes pharmacists, technicians, and any other individuals who assist in the pharmacy.

5. Finally, in some situations, it is necessary for the pharmacy to allow disclosure of PHI to a person or organization that is known under HIPAA as a "business associate." Typically, business associates perform a function that requires disclosure of PHI such as billing services, claims processing, utilization review, or data analysis. Under HIPAA, a pharmacy is allowed to disclose PHI to a business associate if the pharmacy obtains satisfactory assurances, usually in the form of a contract, that the business associate will use the information only for the purposes for which it was engaged by the pharmacy.

HIPAA also provides security provisions. These security provisions went into effect April 20, 2005, almost 2 years after the privacy provisions. The security standards are designed to protect the confidentiality of PHI that is threatened by the possibility of unauthorized access and interception during electronic transmission. Like the privacy provisions, any pharmacy that transmits any health information in electronic form is required to comply with the security rules.

In particular, the security standards define administrative, physical, and technical safeguards that the pharmacist must consider in order to protect the confidentiality, integrity, and availability of PHI.

A unique aspect of the security provisions is that they include both "required and addressable" implementation specifications. Required implementation specifications are those that must be met, whereas, in addressable specifications, the pharmacy must determine whether the suggested safeguards are reasonable and appropriate, given the size and capability of the organization as well as the risk.

While cost may be a factor that a covered entity may consider in determining whether to implement a particular specification, nonetheless a clear requirement exists that adequate security measures be implemented. Cost considerations are not meant to exempt covered entities from this responsibility.

Quick review question 26

Which of the following would not be considered PHI?

   a. Stacy Adams, ovarian cyst
   b. SSN 297-56-4189, peptic ulcer
   c. University Hospital, 13 nosocomial infections
   d. 65 Pololei Drive, 45 y.o. male with HIV

Quick review question 27

When is the first time a pharmacy is required to inform a patient of their privacy practice?

   a. during the first service they provide the patient
   b. not until the patient requests such information
   c. individual pharmacies do not have an obligation to inform patients of their privacy practice
   d. it should be listed in all their advertisements so patients will know it prior to transacting business with them

How long does a new hire in a pharmacy have before they must be trained about the proper use of PHI?

a. during their final interview
b. in an orientation prior to working in the pharmacy
c. within a reasonable time frame after being hired
d. they are expected to have learned it during their schooling and should not need additional training

Quick review question 29

Whom is a pharmacy not allowed to disclose PHI to?

a. a patient's spouse
b. the pharmacy's billing service
c. claims processing
d. utilization review

Drug recalls

Drug recalls are, with a few exceptions, voluntary on the part of the manufacturer. However, once the FDA requests a manufacturer recall a product, the pressure to do so is substantial. The negative publicity from not recalling would significantly damage a reputation, and the FDA could take the manufacturer to court where criminal penalties could be imposed. The FDA can also require recalls in certain instances with infant formulas, biological products, and devices that pose a serious health hazard. Manufacturers may of course recall drugs on their own and do so from time to time for any number of reasons.

Drug recall classifications

There are three drug recall classifications, class I, class II, and class III.

Class I drug recalls have reasonable probability that the use of, or exposure to, a violative product will cause serious adverse health consequences or death. An example would include the diet aid

Fen-Phen (fenfluramine/phentermine) which caused irreparable heart valve damage and pulmonary hypertension.

Class II drug recalls involve the use of, or exposure to, a violative product that may cause temporary or medically reversible adverse health consequences or where the probability of serious adverse health consequences is remote. An example would include a medication that is under-strength, but is not used to treat life-threatening situations.

Class III drug recalls involve medications that the use of, or exposure to, a violative product is not likely to cause adverse health consequences. Examples might be a container defect (plastic material delaminating or a lid that does not seal), off-taste or incorrect color, and simple text errors such as the incorrect expiration date on the manufacturer's bottle.

How drug recalls work

When an FDA-regulated product is either defective or potentially harmful, recalling that product, removing it from the market, or correcting the problem is the most effective means for protecting the public. This is a multi-step process.

Step 1: Reports of adverse effects

FDA first hears about a problem product in several ways:

- A company discovers a problem and contacts the FDA.
- The FDA inspects a manufacturing facility and determines the potential for a recall.
- The FDA receives reports of health problems through various reporting systems.
- The Centers for Disease Control and Prevention (CDC) contacts the FDA.

After receiving enough reports of adverse effects or misbranding, it decides the product is a threat to the public health and it contacts the manufacturer to recommend a recall.

Step 2: Manufacturer agrees to recall

Provided the manufacturer agrees to a recall, they must establish a recall strategy with the FDA that addresses the depth of the recall, the extent of the public warnings, and a means for checking the effectiveness of the recall. The depth of the recall is identified by wholesale, retail, or consumer levels.

Step 3: Customers contacted

Once the strategy is finalized, the manufacturer contacts it customers (pharmacies, wholesalers, etc.) with the following information:

- the product name, size, lot number, code or serial number, and any other important identifying information,
- reasons for the recall and the hazard involved, and
- instructions on what to do with the product, beginning with ceasing distribution.

Step 4: Recall listed publicly

FDA recalls are publicly listed in the FDA's weekly enforcement report (http://www.fda.gov/Safety/Recalls/EnforcementReports/default.htm), see recent recalls at http://www.fda.gov/Safety/Recalls/default.htm , and you can receive email alerts at https://public.govdelivery.com/accounts/USFDA/subscriber/new?topic_id=USFDA_48 . Many subscription services, such as Medscape and Drug Facts and Comparisons eAnswers, also include drug recall information.

Step 5: Effectiveness checks

FDA evaluates whether all reasonable efforts have been made to remove or correct a product. A recall is considered complete after all of the company's corrective actions are reviewed by FDA and deemed appropriate. After a recall is completed, FDA makes sure that the product is destroyed or suitably reconditioned, and investigates why the product was defective in the first place.

Quick review question 30

Which kind of drug recall would involve a medication having reasonable probability of causing death?

a. class A recall
b. class X recall

c. code red
d. class I recall

Quick review question 31

What would be an example of a class II drug recall?

a. a patient wasn't aware of a medication's adverse effects
b. an incorrect expiration date
c. the wrong salt of the drug was utilized
d. the medication is under strength

Quick review question 32

What would be an example of a class III drug recall?

a. a patient wasn't aware of a medication's adverse effects
b. an incorrect expiration date
c. the wrong salt of the drug was utilized
d. the medication is under strength

Quick review question 33

Which organization is responsible for listing drug recalls publicly?

a. DEA
b. FDA
c. TJC
d. USP

Infection control standards

Infection control is concerned with preventing nosocomial or healthcare-associated infections. It is an essential part of the infrastructure of healthcare. Infection control addresses factors related to the spread of infections within the healthcare setting (whether patient to patient, patients to staff, and staff to patient, or among staff), including prevention via hand hygiene, facility and equipment cleaning, and through the use of personal protective equipment (PPE).

Hand hygiene

The Occupational Safety and Health Administration (OSHA) standards require that employers must provide readily accessible hand washing facilities, and must ensure that employees wash hands and any other skin with soap and water or flush mucous membranes with water as soon as feasible after contact with blood or other potentially infectious materials.

The United States Pharmacopeia (USP) chapter 797 states:

"...personnel perform a thorough hand-cleansing procedure by removing debris from under fingernails using a nail cleaner under running warm water followed by vigorous hand and arm washing to the elbows for at least 30 seconds with either nonantimicrobial or antimicrobial soap and water."

Cleaning facilities and equipment

Equipment and facilities for both sterile and nonsterile require cleaning and inspection.

USP 795, with regard to nonsterile compounding states:

"Equipment and accessories used in compounding are to be inspected, maintained, cleaned, and validated at appropriate intervals to ensure the accuracy and reliability of their performance."

Since environmental contact is a common source of contamination with respect to compounded sterile preparations, USP 797 provides a series of standards related to buffer room/clean room, ante room/ante area, and equipment design and cleaning schedules for facilities and equipment used for assembling CSPs. Primary engineering controls (i.e., laminar airflow workbenches, biological safety cabinets, and barrier isolators) which are intimate to the exposure of critical sites, require disinfecting more frequently than do the actual room surfaces such as walls and ceilings.

Cleaning and disinfecting primary engineering controls are the most critical practices before the preparation of CSPs. Such surfaces shall be cleaned and disinfected frequently, including at the beginning of each shift, prior to each batch preparation, every 30 minutes during continuous compounding periods of individual CSPs, when there are spills, and when surface contamination is known or suspected.

With respect to the rooms involved in preparing CSPs, counters and easily cleanable work surfaces must be cleaned daily. Walls, ceilings, and storage shelving must be cleaned on a monthly basis.

Personal protective equipment

When assembling CSPs, preparers should don personal protective equipment (PPE) to further minimize the likelihood that they will contaminate the final product. USP 797 provides requirements for how personnel should wash and garb including the order that they should don their PPE, from dirtiest to cleanest. First, personnel should remove unnecessary outer garments and visible jewelry (scarves, rings, ear rings, etc.). Next, they should don shoe covers, facial hair and hair covers, and masks (optionally they may include a face shield). Next, they must perform appropriate hand hygiene. After hand washing, a nonshedding gown with sleeves that fit snugly around the wrists should be put on. Once inside the buffer area an antiseptic hand cleansing using a waterless alcohol-based hand scrub should be performed. Sterile gloves are the last item donned prior to compounding.

Quick review question 34

Healthcare facilities attempting infection control are trying to prevent the spread of infections between which groups?

  a. patient to patient
  b. patient to staff and staff to patient
  c. among staff
  d. all of the above

Quick review question 35

According to USP 797, what is the minimum amount of time an individual should spend performing hand hygiene prior to assembling CSPs?

  a. there is no minimum
  b. 30 seconds
  c. the time it takes to sing, "Happy Birthday" twice

d. two minutes

Quick review question 36

How often should documented cleanings of the walls and ceilings of a buffer room occur?

a. per shift
b. daily
c. weekly
d. monthly

Quick review question 37

In what order should you don your PPE?

a. from dirtiest to cleanest
b. from cleanest to dirtiest
c. the order doesn't matter
d. pharmacy personnel working in buffer rooms/areas don't require PPE

Professional standards in pharmacy

Pharmacy professionals, including pharmacists, pharmacy technicians, and other pharmacy employees, are expected to actively demonstrate professionalism through attitudes, qualities, and behaviors commensurate to the area of specialized knowledge that they have achieved. The following list provides an overview of these professional standards:

- The first consideration is the health and safety of the patient.
- Honesty and integrity are integral to the high moral and ethical principles of this field.
- Pharmacy technicians are expected to assist pharmacists in providing patients with safe, efficacious, and cost effective access to health resources.
- Technicians and pharmacists should respect the values of each other's abilities as well as those of colleagues and other healthcare professionals.
- Technicians and pharmacists should maintain their competencies and seek to enhance their knowledge and

skills.
- Healthcare professionals need to maintain a patient's rights to dignity and confidentiality.
- Pharmacy professionals may not assist in providing medications and medical devices of poor quality that do not meet the necessary standards established by laws.
- Pharmacy professionals are not to engage in activities that would discredit the profession.
- Pharmacy professionals should engage in and support organizations that promote their professions and the enhancements of those that are in the profession.

Individual state boards of pharmacy (BOP) provide further standards for pharmacy personnel, including job responsibilities and requirements such as registration, licensure, and certification.

Quick review question 38

From the standpoint of professional standards in pharmacy, the first consideration should be?

a. the health and safety of the patient
b. how much revenue the patient can generate
c. your own financial gains
d. how far you live from work

Facility, equipment, and supply requirements

Depending on the specific kind of pharmacy practice, various kinds of facilities, equipments, and supplies will be required. The needs of a hospital pharmacy are different from those of a community pharmacy, just as the needs of a nuclear pharmacy will be different from those of a mail order pharmacy.

All facilities will have to be clean enough to meet any state and local sanitation requirements. Facilities will need to have adequate space to perform their necessary duties. In many states, their Board of Pharmacy (BOP) will establish minimum requirements for space. They will need a properly functioning heating, ventilation, and air-conditioning (HVAC) system to ensure that medications on the shelf are stored at controlled room temperature (15 to 30° C or 59 to 86° F). The facility will need properly maintained and monitored refrigerators (2 to 8° C or 36 to 46° F) and freezers (-25 to -10° C or -13 to 14° F) for medications that need to be stored that way.

Facilities that assemble compounded sterile preparations (CSPs) will need a primary engineering control, such as a laminar airflow workbench, with adequate support systems around it (i.e., a buffer room). According to the United States Pharmacopeia (USP) chapter 797, if a facility prepares more than a low volume of hazardous drugs, it will need a dedicated negative pressure room.

All pharmacies will need ready access to drug information resources. Some BOPs will place very specific requirements on what kind of drug information resources pharmacies need to carry, whereas others may simply require an adequate reference library to meet the needs of the patient population they serve. It is extremely important that whatever drug references are being utilized in a pharmacy, that they are the most up-to-date editions available.

Quick review question 39

What is the acceptable temperature range for a refrigerator in the pharmacy that is intended for the storage of medications?

   a. -25 to -10° C
   b. -13 to 14° C
   c. 2 to 8° C
   d. 59 to 86° C

Quick review question 40

According to the USP, how many hazardous drugs can a facility prepare before requiring a separate negative pressure room?

   a. once a week
   b. upwards of thirty hazardous drugs per month
   c. more than a low volume
   d. one

Answers to quick review questions

1. D - Hazardous drugs are drugs that are known to cause genotoxicity, which is the ability to cause a change or mutation in genetic material; carcinogenicity, the ability to cause cancer in animal models, humans or both; teratogenicity, which is the ability to cause defects on fetal development or fetal malformation; and lastly, hazardous drugs are known to have the potential to cause fertility impairment, which is a major concern for most clinicians.
2. A - Material Safety Data Sheets are required by the Occupational Safety and Health Administration.
3. C - When looking at expirations on medication vials, it is important to note that if a medication only mentions the month and year, but not the day, then you are to treat it as expiring at the end of the month.
4. B - The Joint Commission (TJC requires that internal and external medications must be stored separately. This reduces the potential that someone will dispense or administer an external product for internal use.
5. A - Whenever receiving and/or handling hazardous materials, individuals are recommended to use personal protective equipment (PPE).
6. D - Workers may be exposed to a hazardous drug at any points during its manufacture, transport, distribution, receipt, storage, preparation, and administration, as well as during waste handling and equipment maintenance and repair.
7. C - Partially used bags, vials, syringes, etc. of RCRA listed hazardous pharmaceutical waste MUST be placed in black RCRA approved containers.
8. B - The DEA is the Drug Enforcement Administration.
9. B - Controlled substances are broken into schedules with various prescribing guidelines based on abuse potential, counter balanced by potential medicinal benefit as determined by the Drug Enforcement Administration and individual state legislative branches.
10. C - There are 5 schedules of controlled substances (CI through CV)
11. D - Mobic (meloxicam) is a non-steroidal anti-inflammatory

drug and is not scheduled.

12. A - The only DEA number, of those listed, that could be legitimate for Dr. Ferguson is BF6428521. Option 'B' starts with an unaccepted letter, option 'C' has the wrong check sum value, and option 'D' has too many numbers.

13. C - Oxycontin (oxycodone) is a Schedule II controlled substance and the purchase of Schedule II controlled substances must be authorized by a pharmacist and executed on either a triplicate DEA 222 order form or an electronic 222 form through a controlled substances ordering system (CSOS)

14. A - Restoril (temazepam) is a Schedule IV controlled substance; Schedule III – V drugs may be ordered by a pharmacy or other appropriate dispensary on a general order from a wholesaler. You should check the delivery in against the original order.

15. B - The records involving the transfer of controlled substances must be kept readily available by the pharmacy for two years for inspection by the DEA.

16. B - Tylenol No. 3 (codeine and acetaminophen) is a Schedule III controlled substance. The DEA allows the transfer of original prescription information for Schedule III, IV, and V controlled substances for the purpose of refill dispensing between pharmacies on a one-time basis.

17. D - A pharmacy is required by the DEA to take an inventory of controlled substances every 2 years (biennially).

18. B - There are a number of medications with restrictions due to patient safety concerns. The FDA primarily manages these medications through their Risk Evaluation and Mitigation Strategy (REMS) initiative.

19. C - Thalidomide causes severe and life-threatening birth defects in 40% of infants when taken during pregnancy.

20. D - Once listed in the nonrechallengeable database, they may never use clozapine again.

21. B - Isotretinoin is only available through a restricted program under REMS called iPLEDGE. Prescribers, patients, pharmacies, and distributors must enroll in the iPLEDGE program (https://www.ipledgeprogram.com).

22. D - When a physician qualifies to be allowed to prescribe buprenorphine for addiction, they are given a new DEA

number that begins with 'X'.

23. D - While the primary goal of OBRA-90 was to save the federal government money by improving therapeutic outcomes, the method to achieve these savings was implemented by imposing on the pharmacist counseling obligations, prospective drug utilization review (ProDUR) requirements, and record-keeping mandates.

24. A - As part of the ProDUR, the following are areas for drug therapy problems that the pharmacist must screen for therapeutic duplication, drug–disease contraindications, drug–drug interactions, incorrect drug dosage, incorrect duration of treatment, drug–allergy interactions, and clinical abuse/misuse of medication.

25. C - Under OBRA-90, a physician may counsel a patient about the name and description of the medication; the dosage form, dosage, route of administration, and duration of drug therapy; any special directions and precautions for preparation, administration, and use by the patient; common severe side effects or adverse effects or interactions and therapeutic contraindications that may be encountered; techniques for self-monitoring of drug therapy; proper storage; refill information; and action to be taken in the event of a missed dose.

26. C - PHI is defined as individually identifiable health information transmitted or maintained in any form and via any medium. The 13 nosocomial infections do not identify any particular patient, and would therefore be considered sufficiently deidentified data.

27. A - Their privacy practice notice must be given to every individual no later than the date of the first service provided.

28. C - All employees working in the pharmacy environment in which PHI is maintained must receive training on the regulations within a reasonable time after being hired.

29. A - A pharmacy is not allowed to share PHI with relatives, not even spouses, but they are allowed to share them with partners classified as business associates under HIPAA.

30. D - Class I drug recalls have reasonable probability that the use of, or exposure to, a violative product will cause serious adverse health consequences or death.

31. D - Class II drug recalls involve the use of, or exposure to, a

violative product that may cause temporary or medically reversible adverse health consequences, or where the probability of serious adverse health consequences is remote.

32. B - Class III drug recalls involve medications that the use of, or exposure to, a violative product is not likely to cause adverse health consequences. Examples might be a container defect (plastic material delaminating or a lid that does not seal), off-taste or incorrect color, and simple text errors such as the incorrect expiration date on the manufacturer's bottle.

33. B - FDA recalls are publicly listed in the FDA's weekly enforcement report.

34. D - Infection control addresses factors related to the spread of infections within the healthcare setting whether patient to patient, patient to staff, and staff to patient, or among staff.

35. B - USP 797 states, "...personnel perform a thorough hand-cleansing procedure by removing debris from under fingernails using a nail cleaner under running warm water followed by vigorous hand and arm washing to the elbows for at least 30 seconds with either nonantimicrobial or antimicrobial soap and water."

36. D - Walls, ceilings, and storage shelving must be cleaned on a monthly basis in rooms involved in the preparation of CSPs, including both the buffer room/area and the ante room/area.

37. A - USP 797 provides requirements for how personnel should wash and garb, including the order that they should don their PPE, from dirtiest to cleanest.

38. A - The first consideration is the health and safety of the patient.

39. C - Refrigerated medications should be stored in the range of 2 to 8° C.

40. C - According to the United States Pharmacopeia (USP) chapter 797, if a facility prepares more than a low volume of hazardous drugs, it will need a dedicated negative pressure room.

CHAPTER 3 Sterile and Non-Sterile Compounding

Key concepts

This chapter will cover the following knowledge areas to prepare you for the *Pharmacy Technician Certification Exam*:

- Infection control (e.g., hand washing, PPE)
- Handling and disposal requirements (e.g., receptacles, waste streams)
- Documentation (e.g., batch preparation, compounding record)
- Determine product stability (e.g., beyond use dating, signs of incompatibility)
- Selection and use of equipment and supplies
- Sterile compounding processes
- Non-sterile compounding processes

Terminology

To get started in this chapter, there are some terms that should be defined.

extemporaneous compounding - Extemporaneous compounding can be defined as the preparation, mixing, assembling, packaging, and labeling of a drug product based on a prescription from a licensed practitioner for the individual patient in a form that the drug is not readily available in (extemporaneous = impromptu, compounding = the act of combining things).

United States Pharmacopeia (USP) - The United States Pharmacopeia (USP) is the official pharmacopeia of the United

States, and is published dually with the National Formulary as the USP-NF. Prescriptions and over-the-counter medicines, and other healthcare products sold in the United States, are required to follow the standards in the USP-NF. The USP also sets standards for food ingredients and dietary supplements. Chapters in the USP that are listed as below 1000 are considered enforceable, while chapters enumerated as 1000 or greater are considered guidelines. Therefore, USP 797 and USP 795 are considered enforceable, while USP 1075 and USP 1160 are simply considered guidelines for best practices.

USP 795 - USP Chapter 795, Pharmaceutical Compounding-Nonsterile Preparations, codifies the rules pharmacists and pharmacy technicians must follow when compounding nonsterile formulations intended for humans and animals.

USP 797 - USP Chapter 797, Pharmaceutical Compounding-Sterile Preparations, provides the first set of enforceable sterile compounding standards issued by the United States Pharmacopeia (USP). USP Chapter 797 describes the procedures and requirements for compounding sterile preparations, and sets the standards that apply to all settings in which sterile preparations are compounded.

USP 1075 - USP Chapter 1075, Good Compounding Practices, is intended to provide guidelines on applying best practices in compounding, both sterile and nonsterile.

USP 1160 - USP Chapter 1160, Pharmaceutical Calculations in Prescription Compounding, provides general information on the mathematical concepts required for compounding pharmaceutical preparations.

The Joint Commission (TJC) - The Joint Commission, formerly the Joint Commission on Accreditation of Healthcare Organizations (JCAHO), is a nonprofit organization that accredits more than 20,000 healthcare organizations and programs in the United States.

Board of Pharmacy (BOP) - Each state has its own board of pharmacy. The BOP sets standards, roles, and requirements for pharmacy personnel and practice setting in their state.

Occupational Safety and Health Administration (OSHA) - OSHA is a government agency within the United States Department of Labor responsible for maintaining safe and healthy work environments.

Material Safety Data Sheet (MSDS) - OSHA-required notices on hazardous substances which provide hazard, handling, clean-up, and first aid information.

personal protective equipment (PPE) - Personal protective equipment is worn by an individual to provide both protection to the wearer from the environment or specific items they are manipulating, and to prevent exposing the environment or the items being manipulated directly to the wearer of the PPE.

nosocomial infection - A nosocomial infection is an infection acquired while in the hospital.

batch preparation - A batch preparation is one in which multiple identical units are prepared in a single operation in anticipation of prescriptions.

trituration - Trituration is a method to reduce particle size (comminution); it may also include the grinding together of two or more substances in a mortar to mix them, as you should want a fine powder to make incorporation better. Trituration is achieved by firmly holding the pestle and exerting a downward pressure with it while moving it in successively larger circles starting at the center of the mortar, moving outward to the side of the mortar, then back again toward the center.

spatulation - Spatulation is the mixing of powders and semi-solids (ointments, creams, etc.) on an ointment pad or slab using a spatula. With this method, there is no particle size reduction, so the powders to be mixed must be fine and of uniform size.

geometric dilution - Geometric dilution, also called geometric incorporation, is the process by which a homogeneous mixture of even distribution of two or more substances is achieved. When using this method, the smallest quantity of active ingredient is mixed thoroughly with a proportion quantity of the diluent or base on the ointment slab. More diluent (base) is added in amounts proportionate to the volume of the mixture on the ointment slab. This process is repeated until all of the ingredients are incorporated in the mixture.

levigation - Levigation is the process of reducing particle size of a solid by triturating it in a mortar or spatulating it on an ointment slab or pad with a small amount of a liquid called a levigating agent. Levigating agents make incorporating solids easier, and they make a smooth, elegant preparation.

parenteral - A parenteral can be defined as any administration route not involving the GI tract, but it is more commonly used in reference to routes that require sterile preparations (injectable routes, ophthalmic, and inhalation).

compounded sterile preparations (CSP) - Compounded sterile preparations are admixtures that need to be assembled under aseptic conditions to prevent contamination.

aseptic techniques - Aseptic techniques are techniques or methods that maintain the sterile condition of products.

pyrogens - Pyrogens are chemicals produced by microorganisms that can cause pyretic (fever) reactions in patients.

precipitate - A precipitate is an insoluble substance separated from a solution due to a reaction between incompatible substances.

osmolarity - Osmolarity, also called osmotic pressure, is a characteristic of a solution determined by the number of dissolved particles in it.

isotonic - An isotonic solution has an osmolarity equivalent to that of blood.

hypertonic - A hypertonic solution has a greater osmolarity than that of blood.

hypotonic - A hypotonic solution has a lesser osmolarity than that of blood.

ions - Ions are molecular particles that carry electric charges.

additive - An additive, in the context of sterile compounding, is a drug that is added to a parenteral solution.

admixture - An admixture is the resulting solution when a drug is added to a parenteral solution.

lyophilized - Lyophilized powders are simply freeze-dried powders.

diluent - A diluent is a solvent that dissolves a lyophilized powder or dilutes a solution.

coring - Coring occurs when a needle damages the rubber closure of a parenteral container causing fragments of the closure to fall into the container and contaminate its contents.

HEPA filter - A HEPA filter is a high efficiency particulate air filter.

laminar airflow - Laminar airflow provides continuous air movement at a uniform rate in one direction.

ISO Class - The International Organization for Standardization has established various levels of air cleanliness. The lower the number, the fewer particles that are suspended in it, and the cleaner the air is. The ISO classes commonly discussed in sterile compounding are ISO Class 5 (3,520 particles of 0.5m micron or larger in a cubic meter), ISO Class 7 (352,000 particles of 0.5m micron or larger in a cubic meter), and ISO Class 8 (3,520,000 particles of 0.5m micron or larger in a cubic meter).

parenteral nutrition (PN) - Parenteral nutrition is a complex solution with two base solutions (amino acids and dextrose) and additional micronutrients. Lipids may or may not also be present in parenteral nutrition.

automated compounding device (ACD) - An automated compounding device is a machine used to prepare CSPs (e.g., PN).

hazardous drugs - Hazardous drugs are drugs that are known to cause genotoxicity, which is the ability to cause a change or mutation in genetic material; carcinogenicity, which is the ability to cause cancer in animal models, humans or both; teratogenicity, which is the ability to cause defects on fetal development or fetal malformation; and lastly, hazardous drugs are known to have the potential to cause fertility impairment, which is a major concern for most clinicians. These drugs can be classified as antineoplastics, cytotoxic agents, biologic agents, antiviral agents, and immunosuppressive agents.

antineoplastic agents - Antineoplastic agents are drugs that inhibit and combat the development of tumors.

ante area - An area that is ISO class 8 or better where hand hygiene and garbing is performed. It also functions as a transition area between the buffer area and the rest of the facility.

buffer area - An area that is required to be ISO class 7 or better. A primary engineering control (PEC), such as a laminar airflow workbench, would be located in this area.

primary engineering control (PEC) - A primary engineering control, which could include a room or device, provides an ISO Class 5 environment for compounding sterile products. Examples of primary engineering control may include laminar airflow workbenches (LAFW), compounding aseptic isolators (CAI), biological safety cabinets (BSC), compounding aseptic containment isolators (CACI), and clean rooms that create an ISO Class 5 environment.

positive pressure room - A room that has higher pressure than adjacent spaces and therefore has a net flow of air out of the room.

negative pressure room - If a pharmacy prepares more than a low volume of hazardous drugs, the appropriate type of PEC should be located within a negative pressure room. A negative pressure room has a lower pressure than adjacent rooms and therefore the net flow of air is into the room.

Quick review question 1

Which chapters in the USP are considered enforceable?

    a. all the chapters below 1000
    b. all the chapters numbered 1000 or greater
    c. all of the chapters
    d. none of the chapters

Quick review question 2

Which chapter in the USP establishes enforceable regulations for sterile compounding?

    a. USP 795
    b. USP 797
    c. USP 1075

d. USP 1160

Quick review question 3

Where is a nosocomial infection acquired?

a. at home
b. in the community
c. in a hospital
d. from your pets

Secundum artem

Extemporaneous compounding, often just called compounding, is a necessary skill for many pharmacists and pharmacy technicians. Many compounders will use the Latin term *secundum artem* (according to the art) when referring to the skills required to prepare these products. Extemporaneous compounding can be defined as the preparation, mixing, assembling, packaging, and labeling of a drug product based on a prescription from a licensed practitioner for the individual patient in a form that the drug is not readily available (extemporaneous = impromptu, compounding = the act of combining things). Extemporaneous compounding is required for prescription orders that are not commercially available in the requested strength or dosage form.

There are two types of pharmaceutical compounding: sterile (ophthalmics, IVs, parenteral nutrition, chemotherapy, and various other injectables) and nonsterile (creams, ointments, oral suspensions, capsules, suppositories, medication sticks, troches, etc.).

Quick review question 4

What are the two types of compounding?

a. professional and amateur
b. pharmacist and technician
c. institutional and community
d. sterile and nonsterile

Basic steps for preparing compounded drug preparations

Regardless of the compounding category, there is a general order of work for all extemporaneous compounding. By following a procedure, you will be able to work more effectively.

1. Carefully read and interpret the prescription or medication order.
2. Note any missing or confusing information; clarify, gather, and add this information to the drug order.
3. Check the dose,dosage regimen, dosage form, and route of administration for appropriateness.
4. Determine a preliminary compounding procedure.
5. Perform necessary calculations. If possible, have a colleague check calculations.
6. Select required ingredients. It is the responsibility of the pharmacist to choose appropriate quality ingredients for compounding. The preferred grade for this is USP or NF. If an official USP or NF ingredient is not available, the pharmacist should use professional judgment in the selection of an alternative source so that the safety and purity of the ingredient is assured. This may require requesting a certificate of analysis from the supplier.
7. Choose appropriate compounding equipment.
8. Using recommended techniques, prepare the product.
9. A visual inspection of the product should be done.
10. Choose an appropriate container and package the preparation.
11. Determine an appropriate beyond use date.
12. Label container and include recommended auxiliary labels.
13. Document the compounding process.

Safety

Extemporaneous compounding brings with it its own safety concerns. There is a vast array of information to be aware of with safety requirements depending as to what kinds of chemicals you are working with. Of particular concern is infection control through proper hand hygiene and personal protective equipment, and proper handling with regard to disposal and waste management.

Infection control

Infection control is concerned with preventing nosocomial or healthcare-associated infections. Infection control addresses factors related to the spread of infections within the healthcare setting (whether from patient to patient, from patient to staff, from staff to patients, or among staff), including prevention via hand hygiene and through the use of personal protective equipment (PPE).

When assembling CSPs, preparers should don personal protective equipment (PPE) to further minimize the likelihood that they will contaminate the final product. USP 797 provides requirements for how personnel should wash and garb, including the order that they should don their PPE, from dirtiest to cleanest. First, personnel should remove unnecessary outer garments and visible jewelry (scarves, rings, ear rings, etc.). Next, they should don shoe covers, facial hair and hair covers, and masks (optionally they may include a face shield). Next, they must perform appropriate hand hygiene.

USP 797 states:

"...personnel perform a thorough hand-cleansing procedure by removing debris from under fingernails using a nail cleaner under running warm water followed by vigorous hand and arm washing to the elbows for at least 30 seconds with either nonantimicrobial or antimicrobial soap and water."

After hand washing, a nonshedding gown with sleeves that fit snugly around the wrists should be put on. Once inside the buffer area, an antiseptic hand cleansing using a waterless alcohol-based hand scrub should be performed. Sterile gloves are the last item donned prior to compounding.

Disposal

Proper waste stream management is very important in pharmacy as mismanagement can have negative impacts on the community, the environment, and can cause fines to be levied against a pharmacy or health care facility. Waste that poses no ecological threats, is devoid of protected heath information, and poses no major safety risks may be disposed of in regular waste receptacles.

Broken ampules, syringes, and various needles need to be treated with caution to avoid accidental needle sticks and/or cuts. If the medications or diluents that these sharps were exposed to are not classified as hazardous, then you may dispose of this waste in red sharps containers.

Properly labeled, leak-proof, and spill-proof containers of nonreactive plastic are required for areas where hazardous waste is generated, and can be further broken down into either yellow or black containers. Fully used vials, syringes, tubing, and bags of hazardous drug waste, along with PPE used while working with hazardous drugs, may be disposed in yellow, properly labeled containers. Also, any partially used or expired hazardous drugs that are not also considered to be RCRA (Resource Conservation and Recovery Act) regulated hazardous drugs should be placed in these yellow buckets. Trace contaminated items such as booties, gowns, and masks, even if they were not involved in spills, MUST be treated as hazardous waste (the yellow buckets are sufficient for this).

Partially used bags, vials, syringes, etc. of RCRA listed hazardous pharmaceutical waste MUST be placed in black RCRA approved containers. Hazardous waste must be properly manifested and transported by a federally permitted hazardous waste transporter to a federally permitted hazardous waste storage, treatment, or disposal facility.

Quick review question 5

Which USP chapter details the proper use of PPE?

  a. 795
  b. 797
  c. 1075
  d. 1160

Quick review question 6

What kind of container should waste that poses no ecological threats, is devoid of protected heath information, and poses no major safety risks be disposed of in?

a. red
b. yellow
c. black
d. in the general trash, as long as it has been deidentified

Quick review question 7

Broken ampules, syringes, and various needles need to be treated with caution to avoid accidental needle sticks and/or cuts. If the medications or diluents that these sharps were exposed to are not classified as hazardous, what color containers should they be disposed of in?

a. red
b. yellow
c. black
d. in the general trash, as long as it has been deidentified

Quick review question 8

What color containers should fully used vials, syringes, tubing, and bags of hazardous drug waste, along with PPE used while working with hazardous drugs be disposed in?

a. red
b. yellow
c. black
d. in the general trash, as long as it has been deidentified

Documentation

Documentation of extemporaneously compounded medications is an important responsibility within the pharmacy as it provides a chain of accountability for the preparations and it aids in tracking drug recalls. Documentation can be broken into two major categories: record keeping and labeling.

Record keeping

Whenever preparing any extemporaneous compound, the minimum

information to maintain would be a method for tracking whom the compounder(s) is/are, who verified that the procedures for compounding were carried out correctly, and drugs/chemicals used including manufacturer, lot number, expiration date, and an internally assigned batch number and beyond-use date. Many facilities will also track information on all the equipment used in preparing the compound.

Sometimes facilities will compound batch preparations. A batch preparation is one in which multiple identical units are prepared in a single operation in anticipation of prescriptions. Some examples of batch preparations might include preparing 100 capsules for hormone replacement therapy when only 30 have been ordered, or preparing 10 IV minibags of the same drug at the same concentration from a bulk container when only two have been ordered. These preparations are being made in anticipation of additional orders being received in the pharmacy. Since none of these medications will be dispensed or sold without a specific patient in mind, it does not cross the line from compounding to manufacturing. Documentation must also be maintained on all of these batch preparations as well.

Labeling

An important aspect of documentation is providing proper labeling of compounded preparations for others to read once dispensed from the pharmacy. While the requirements and expectations of what kind of labeling information should be present on a medication dispensed to a patient for use at home will vary from those intended for use in a institutional setting there are some expectations and requirements of both. The active ingredients and concentrations should be present as well as a tracking number (whether a batch number or prescription number), a verification of who prepared and checked the preparation (even if just listed by initials), and an appropriate beyond-use date.

Quick review question 9

Which of the following should be included in the documentation of a compounded prescription?

a. the chemicals used in the preparation and their source
b. who compounded the preparation
c. a beyond-use date
d. all of the above

Quick review question 10

Which of the following statements about a batch preparation is true?

a. A pharmacy should have a manufacturing license prior to compounding a batch preparation.
b. A batch preparation involves preparing multiple of an identical unit.
c. A batch preparation may not be done in anticipation of use.
d. None of the above statements about batch preparations are true.

Product stability

A compounded medication needs to be stable for its intended purpose. This includes both concerns over chemical compatibility and labeling the product with an appropriate beyond-use date.

Chemical compatibility

Chemical compatibility is a measure of how stable a substance is when mixed with another substance. Some incompatibilities are visible (color change or formation of a precipitate); other chemical incompatibilities can be equally concerning even if not visible, such as the formation of gases and volatile chemicals. An example of a chemical incompatibility is the formation of a calcium phosphate precipitate that falls out of solution when calcium gluconate and potassium phosphate are mixed together in sufficient concentration.

Beyond-use date

The requirement to provide a beyond-use date sometimes causes confusion as there is some ambiguity between what an expiration date is and what a beyond-use date is. An expiration date is the date put on the label of a drug product by the manufacturer or distributor of the product. Federal law requires that manufactured (compounding is not considered manufacturing) drug products be labeled with an expiration date. The beyond-use date is the date placed on a prescription bottle by a pharmacy noting when that prescription should no longer be used.

The American Medical Association (AMA) states:

"The beyond-use date placed on the label shall be no later than the expiration date on the manufacturer's container. The beyond-use date is a date after which an article [drug] must not be used. Based on the information supplied by the manufacturer, the dispenser shall place on the label of the prescription container a suitable beyond-use date to limit the patient's use of the article."

For water containing formulations prepared from ingredients in solid form, the beyond use date is not later than 14 days when stored at cold temperatures. Generally, a two-week dating is the maximum for ointments containing water if a preservative is not present. Ointment with water are subject to microbial growth. Ointments without water can have much longer dating. Ointment jars, although widely used, expose the preparation to air and microbial contamination when opened and when the ointment is removed using the fingers. The maximum beyond-use date for a very stable preparation would be 6 months or 25% of the time remaining between the time of compounding and the shortest expiration date of the ingredients, whichever is earlier. You need to consult available literature, if possible, and professional judgment is required in these cases. For all other formulations, the beyond-use date is not later than the intended duration of therapy or 30 days, whichever is earlier.

Quick review question 11

Which of the following could be caused by a chemical incompatibility between drugs?

    a. a precipitate
    b. a color change
    c. the formation of a gas
    d. all of the above

Quick review question 12

Which of the following statements about beyond use dates is

considered most accurate?

a. A beyond-use date is the same thing as an expiration date.
b. A beyond-use date may not exceed the expiration date.
c. All beyond-use dates can be found in the USP.
d. all of the above

Nonsterile compounding

Compounding pharmacies are capable of preparing a broad array of dosage forms including capsules, tablets, oral suspensions and solutions, suppositories, troches (lozenges), various topical dosage forms (gels, lotions, ointments, creams), lollipops, and more. One of the things that tends to make compounding both interesting and difficult is finding the best way to manipulate a dosage form to benefit a patient.

USP Chapter 795, Pharmaceutical Compounding-Nonsterile Preparations, codifies the rules pharmacists and pharmacy technicians must follow when compounding nonsterile formulations intended for humans and animals. This chapter provides minimal standards for equipment, supplies, labeling, beyond-use dates, compounding processes, and definitions of various dosage forms.

USP 1075 has designated three categories of compounding related to different levels of required experience, training, and physical facilities. The considerations factoring into this are:

- the difficulty/complexity of the compound,
- product stability and warnings,
- packaging and storage requirements,
- the physical dosage forms,
- the complexity of calculations required,
- local versus systemic effects of the product,
- risk to the compounder, and
- risk to the patient.

These categories are broken into Simple, Moderate, and Complex compounding. Simple compounding involves preparing compounds from USP monographs and peer-reviewed journals or even just reconstituting and manipulating commercial products within manufacturer guidelines. Required procedures and equipment are well documented. Moderate compounding requires special calculations and procedures, and is generally not as well documented as Simple compounding. Complex compounding requires special training, environment, facilities, equipment, and procedures.

Equipment and supplies

The following is a list of equipment commonly used for nonsterile compounding.

- Mortar and pestle

- Electronic balance

- Class A prescription balance (Class III balance)

- Analytical balance

- Pharmacy weights

- Glassine papers (weighing papers)

- Weighing boats (weighing canoes)

- Ointment slab or pad

- Spatula

- Spatula cards

- Scoopula

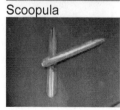

- Glass stirring rods

- Hot plate

- Magnetic stir

- Beakers

- Graduated cylinder

- Erlenmeyer flask

- Droppers

- Funnels

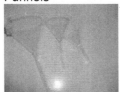

- Oral syringes

- Ointment jars and tubes

- Prescription vials

- Bottles

Processes

Accuracy is a chief concern when it comes to weighing (pharmacy prefers the term weight to mass) and measuring the necessary quantity of ingredients for pharmaceutical compounding. The United States Pharmacopeia (USP) allows a tolerance of plus or minus 5 percent for most preparations.

Weighing

Weighing items may involve a traditional prescription balance along with a set of weights, or by more modern means, using an electronic and/or analytical balance.

A Class A prescription balance is also known as a double-pan (2-pan) torsion balance (officially designated as a Class III balance). Class III balances have a minimal weighable quantity (MWQ) of 120 mg, and must have a maximum weighable quantity (the capacity) of 60 grams, though some will weigh up to 120 grams; you need to look at the stated capacity of the balance.

The sensitivity (sensitivity requirement) of a balance is the amount of weight that will move the balance pointer one-division marker on the marker plate. If you attempt to weigh something less than the balance's sensitivity requirement, the balance will give you a reading of zero (that is, the mass on the balance is too small for it to sense).

A Class III balance has a maximum sensitivity requirement of 6 mg to produce a change of one index division in the rest point as shown by the balance indicator. The smaller the weight required to produce this one division displacement, the greater is the sensitivity of the balance.

Electronic single-pan balances have internal weights, digital display features, and readability and precision of 1 mg, and are available at relatively reasonable costs. Most find these balances easier to use and more accurate than a traditional double-pan balance.

Never place weights, drugs, or chemicals directly onto a balance pan. This includes the balance pan of an electronic or analytical balance. Glassine weighing papers or weighing canoes are generally preferred for weighing. Glassine papers have a smooth, shiny surface that does not absorb materials placed on them, and drugs and chemicals are easily slipped off for a complete transfer. The paper should be diagonally creased from each corner or folded in quarters, and then flattened and placed on the pans. This helps to contain the substance being weighed and prevents spilling on the balance pans or balance platform. New weighing papers or weighing boats should be used with each new drug to prevent contamination.

There are some general rules about using a balance that help to maintain a prescription balance in top condition.

- Always use the balance on a level surface and in a draft free area.
- Always cover both pans with weighing papers or use

weighing boats. These protect the pans from abrasions, eliminate the need for repeated washing, and reduce loss of drugs to porous surfaces.

- A clean paper or boat should be used for each new ingredient to prevent cross contamination of components.
- The balance must be readjusted after a new weighing paper or boat has been placed on each pan. Weighing papers taken from the same box can vary in weight by as much as 65 mg. Larger weighing boats can vary as much as 200 mg.
- Always arrest the balance before adding or removing weight from either pan. Although the balance is noted for its durability, repeated jarring of the balance will ultimately damage the working mechanism of the balance and reduce its accuracy.
- Use a spatula to add or remove ingredients from the balance. Do not pour ingredients out of the bottle.
- Always clean the balance, close the lid, and arrest the pans before storing the balance between uses.

Electronic and analytical balances have digital displays and may have internal calibration capabilities. These may either be top-loading or encased to protect the balance from dust and drafts. Electronic balances generally have a greater level of accuracy than a prescription balance.

An analytical balance is used to measure mass to a very high degree of precision and accuracy. The measuring pan(s) of a high precision (0.1 mg or better) analytical balance are inside a transparent enclosure with doors so that dust does not collect and so any air currents in the room do not affect the balance's operation. An analytical balance is usually ten-fold more accurate than an electronic balance.

To use either an electronic or analytical balance, follow these guidelines:

- Always use the balance on a level surface in a draft free area.
- If the balance has a level bubble, make sure the bubble is inside the bulls eye and make adjustments in the leveling feet as needed.

- Place a weighing boat or a single sheet of weighing paper on the pan.
- When the balance has determined the final weight, press the tare bar to compensate for the weight of the weighing boat or weighing paper.
- As ingredients are added or removed from the weighing boat, the digital display will show the weight of the ingredient in the boat.
- Make sure the balance has determined the final weight before adding or removing any ingredient.

Measuring

Graduates (both cylinders and conical), volumetric flasks (such as Erlenmeyer flasks and beakers), burettes, pipettes, medicine droppers, and syringes are examples of volumetric equipment. The term volumetric simply means measures volume; therefore, volumetric equipment is used for measuring volumes of liquid. Volumetric vessels are either to deliver (TD) or to contain (TC). To deliver means that the vessel must be completely emptied to dispense the needed volume. Single volume pipettes, syringes, droppers and small calibrated pipettes are TD vessels. To contain means that the vessel does not need to be completely emptied to dispense the needed volume. Volumetric flasks, graduates, and some calibrated pipettes are TC vessels.

For maximum accuracy in measuring, select a measuring device with a capacity equal to or slightly larger than the volume to be measured. As graduated cylinders are usually more accurate than conical graduates or Erlenmeyer flasks, we will focus on them. The general rule is to measure volumes no less than 20% of the capacity of a graduate. For example, 2 mL would be the minimum measurable quantity (MMQ) for a 10 mL graduate. Whenever you are measuring items in a graduated cylinder, you should look at the bottom of the meniscus. The meniscus is the curve in the upper surface of a liquid close to the surface of the container or another object caused by surface tension. Water and oil based substances typically have a concave meniscus.

Quick review question 13

What is the MWQ of a class III prescription balance?

a. 0.1 mg
b. 1 mg
c. 120 mg
d. 60 g

Quick review question 14

When using an analytical balance, what is the best practice for compensating for the weight of the weighing paper placed on the balance pan?

a. The weight is negligible and can be ignored.
b. Write down the weight of the paper on your compounding form and subtract it from your ingredient weight.
c. Once the balance has determined the weight of the paper, press the tare bar.
d. None of the above are the correct answers.

Quick review question 15

What is the general rule as to how small a volume can be accurately measured in a graduated cylinder?

a. 1 mL
b. It depends on what the smallest marking is on the graduated cylinder.
c. No less than 20% of the capacity of a graduate
d. None of the above are the correct answers.

Quick review question 16

What kind of dosage forms can compounding pharmacies most commonly prepare?

a. solid dosage forms - capsules, troches, lollipops
b. topical dosage forms - gels, lotions, ointments, creams
c. oral liquids - solutions, suspensions
d. all of the above

Sterile compounding

Many facilities find it necessary to provide patient specific compounded sterile preparations including hospitals, surgicenters, ambulatory care facilities, long term care facilities, home infusion companies, nuclear pharmacies, etc. USP Chapter 797, Pharmaceutical Compounding-Sterile Preparations, provides the first set of enforceable sterile compounding standards issued by the United States Pharmacopeia (USP). USP Chapter 797 describes the procedures and requirements for compounding sterile preparations, and sets the standards that apply to all settings in which sterile preparations are compounded.

USP 797 is based around three categories of microbial contamination: low-risk level CSPs, medium-risk level CSPs, and high-risk level CSPs. It is important to remember that these risk categories are based solely on contamination risk, and do not include any consideration of potential risks to the compounder. There are also additional requirements for compounding hazardous drugs.

Low-risk compounding is performed within an ISO Class 5 (or better) environment and involves the use of no more than three sterile products and no more than two entries through a medication port.

Medium-risk compounding is performed within an ISO Class 5 (or better) environment and involves more complex manipulations of sterile products, including more product manipulations and potentially more medication port entries than Low-risk compounding. As long as all products and equipment used directly in compounding are considered sterile to begin with, it is still considered medium-risk. A facility designed to prepare medium-risk compounds may also prepare low-risk compounds.

High-risk compounding is performed within an ISO Class 5 (or better) environment, but involves the use of either nonsterile products and/or equipment. These medications must be terminally sterilized prior to use for a patient. Terminal sterilization methods may include filtration through a 0.2 micron filter, the use of an autoclave, and/or gamma irradiation. A facility designed to prepare high-risk compounds may also prepare low and medium-risk compounds.

Hazardous drugs provide additional safety concerns for the preparer. A facility that prepares more than a low volume of hazardous drugs is required to have a separate negative pressure room in which an appropriate primary engineering control will be housed. Primary engineering controls used to prepare hazardous drugs should not be used to prepare non-hazardous drugs.

Facilities

A facility must have an adequate design to support the risk level of compounding performed within it. The following two images are examples of potential layouts for low and medium-risk level compounding compared to high-risk level compounding and hazardous drug compounding.

EXAMPLE OF CLEAN ROOM FLOOR PLAN SUITABLE FOR LOW AND MEDIUM-RISK LEVEL CSPs

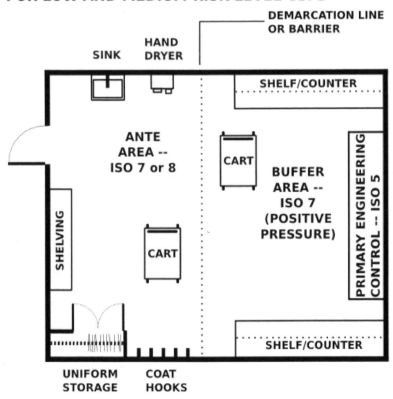

Pharmacy personnel will enter through the ante area where they will perform appropriate cleansing and garbing. Non-hazardous medications will then be taken into a positive pressure buffer area and prepared within a primary engineering control, such as a laminar airflow workbench (LAFW) or a compounding aseptic isolator (CAI).

EXAMPLE OF CLEAN ROOM FLOOR PLAN SUITABLE FOR HIGH-RISK LEVEL CSPs AND HAZARDOUS DRUGS

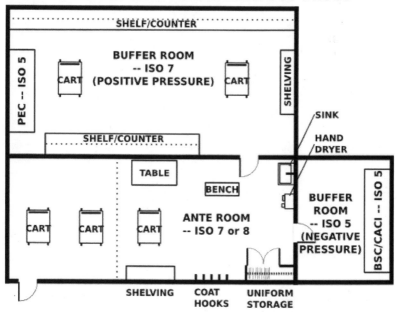

If a pharmacy prepares high-risk compounds, the buffer room and the ante room need to be separate rooms.

If preparing more than a low volume of hazardous drugs, pharmacies will need a separate negative pressure buffer area, which will house a primary engineering control, such as a biological safety cabinet (BSC) or a compounding aseptic containment isolator (CACI).

Equipment and supplies

Below is a list of equipment commonly used to prepare CSPs.

- Laminar airflow workbench (LAFW)

- Compounding aseptic isolator (CAI)

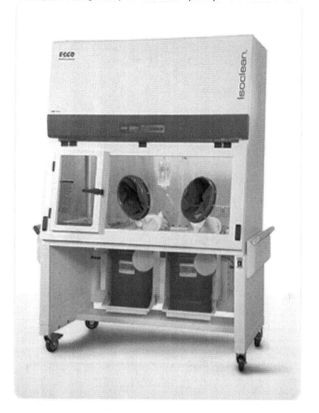

- Biological safety cabinet (BSC)

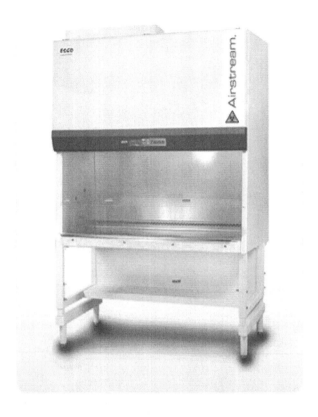

- Compounding aseptic containment isolator (CACI)

- Ampules

- Single dose vial

- Multi-dose vial

- Sterile powder for injection

- Prefilled disposable syringes

- Ready-to-mix systems (Add-Vantage, Minibag Plus, etc.)

- Dual chamber vial

- Minibags

- Large volume parenteral

- Glass bottle for IV fluids

152

- Duplex bags

- Administration sets

- Needles

- Syringes

- Filter needles

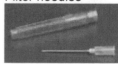

Processes

Accuracy of measurement is a primary concern when performing sterile compounding. Measurements are primarily performed using syringes. Depending on the size of the syringe, it may have markings as small as 0.01 mL on a 1 cc syringe (commonly called a tuberculin syringe), whereas a 60 mL syringe may only have a marking every 1 mL. This is why it is important to use a syringe that comes closest to the volume that needs to be measured, which ideally can also contain the full quantity desired. A 60 cc syringe would not be used to measure 0.25 mL, and a 1 cc syringe should not be used to measure 50 mL. Common syringe sizes include 1 cc, 3 cc, 5 cc, 10 cc, 20 cc, 30 cc, and 60 cc.

Quick review question 17

If a parenteral nutrition is prepared and involves 17 sterile products being combined within an appropriate PEC, which risk level would this compound be considered?

a. low-risk
b. medium-risk
c. high-risk
d. hazardous

Quick review question 18

A facility that compounds more than a low volume of hazardous drugs should prepare hazardous drugs in what kind of room?

a. negative pressure room
b. neutral pressure room
c. positive pressure room
d. none of the above

Quick review question 19

Which of the following is not an example of an appropriate PEC?

a. an ISO Class 5 clean room
b. LAFW
c. BCS
d. CAI

Quick review question 20

Which syringe size would be most appropriate to use if you required 5.7 mL of solution?

a. tuberculin syringe
b. 5 cc syringe
c. 10 cc syringe
d. 60 cc syringe

Quick review question 21

If a CSP requiring the use of a nonsterile powder is prepared within an appropriate PEC, and then terminally sterilized in an autoclave, which risk level would this compound be considered?

a. low-risk
b. medium-risk
c. high-risk
d. hazardous

Calculations often used for compounding

Pharmacy compounding often requires pharmacists and technicians to perform various calculations in order to prepare a final product. In the following section we will review common conversions, common problem solving methods, such as ratio proportions and dimensional analysis, and look at various scenarios involving compounding and calculations.

As this chapter section is only intended as review, if you want a longer explanation about these calculation concepts look at *USP 1160 Pharmaceutical Calculations in Prescription Compounding* (this chapter can be found online via a search engine) or for an even more in depth explanation with numerous practice problems, look at *Pharmaceutical Calculations* by Sean Parsons (this can be downloaded for free at http://pharmaceuticalcalculations.org/).

Conversions

With respect to conversions, it makes most sense to review the English (or household) system, the metric system, and common conversions between the two.

English/household system

Weight

16 ounces (oz) = 1 pound (lb)

Volume

3 teaspoons (tsp) = 1 tablespoon (Tbs)
2 tablespoons (Tbs) = 1 fluid ounce (fl oz)
8 fluid ounces (fl oz) = 1 cup
16 fluid ounces (fl oz) = 1 pint (pt)

Length

12 inches (in) = 1 foot

Metric system

Weight

1000 nanograms (ng) = 1 microgram (mcg)
1000 micrograms (mcg) = 1 milligram (mg)
1000 milligrams (mg) = 1 gram (g)
1000 grams (g) = 1 kilogram (kg)

Volume

1 cubic centimeter (cc) = 1 milliliter (mL)
1000 milliliters (mL) = 1 liter (L)

Length

1000 millimeters (mm) = 1 meter (m)
100 centimeters (cm) = 1 meter (m)

Conversions between English/household system and metric system

Weight

1 ounce (oz) = 28.4 grams (g)
1 pound (lb) = 454 grams (g)
1 kilogram (kg) = 2.2 pounds (lbs)

Volume

1 teaspoon (tsp) = 5 milliliters (mL)
1 tablespoon (Tbs) = 15 milliliters (mL)
1 fluid ounce (fl oz) = 30 milliliters (mL)
1 pint (pt) = 480 milliliters (mL)

Length

1 inch (in) = 2.54 centimeters (cm)

Ratio proportion

A ratio proportion is a statement of equality between two ratios. An example would be:

$$\frac{1}{2} = \frac{2}{4}$$

Since the ratios are equal, if there is a missing piece of data (a variable that needs to be solved for), you can cross multiply and divide to find the value of the missing data. The labels also cross multiply and cancel out as appropriate. When cross multiplying to solve for a variable, multiply the numbers that are diagonal to each other and then divide by the number diagonal to the variable.

Example: Magnesium sulfate is available as 5 grams per 10

milliliters, so how many milliliters will you need for a dose of 2 grams?

$$\frac{5\,g}{10\,mL}=\frac{2\,g}{N}$$

$$10\,mL \times 2\,g \div 5\,g = \mathbf{4\,mL}$$

Notice that the grams canceled each other out since they divided into each other.

Dimensional analysis

Dimensional analysis (also called factor label) involves multiplying a series of fractions in such a way that units are canceled out till you are left for the units you are looking for.

Example: To demonstrate this concept we will look at how many fluid ounces are in 2 liters:

$$\frac{2\,L}{1} \times \frac{1000\,mL}{1\,L} \times \frac{1\,fl\,oz}{30\,mL} = \mathbf{66.7\,fl\,oz}$$

Percentage strength

There are many ways to express the concentration of a drug; one of the most common is percentage strength. There are three kinds of percentage strength that you will frequently use when doing dosage calculations:

weight/weight percent (w/w%) - typically measured in grams/100 grams and include items like ointments and creams
volume/volume percent (v/v%) - typically measured in milliliters/100 milliliters and include items like alcohol preparations and oil-in-water preparations
weight/volume percent (w/v%) - measured in grams/100 milliliters and includes drugs dissolved in solution

An additional item to help you solve percentage strength problems when setting them up as ratio proportions is to think of it as *'active ingredient over mixture'*.

The following are some examples of percentage strength:

Example: What would be the weight, expressed in grams, of zinc oxide (zinc oxide is the active ingredient) in 120 grams of a 10% zinc oxide ointment (the ointment is the total mixture)?

$$\frac{10 \, g \, ZnO}{100 \, g \, ung} = \frac{N}{120 \, g \, ung}$$

$N = 12 \, g \, \textbf{Zinc Oxide}$

Example: How many milliliters of isopropyl alcohol are in a 480 milliliter bottle of 70% isopropyl alcohol?

$$\frac{70 \, mL \, isopropyl \, alcohol}{100 \, mL \, sol} = \frac{N}{480 \, mL \, sol}$$

$N = 336 \, mL \, \textbf{isopropyl alcohol}$

Example: If 4.5 grams of a medication is dissolved in 2 fluid ounces of solution, what is the percentage strength? You will need to convert the fluid ounces to milliliters before solving this problem.

$$\frac{2 \, fl \, oz}{1} \times \frac{30 \, mL}{1 \, fl \, oz} = 60 \, mL$$

$$\frac{4.5 \, g}{60 \, mL} = \frac{N}{100 \, mL}$$

$N = 7.5 \%$

Diluting stock solutions

A stock solution is a concentrated solution from which less concentrated solutions may be prepared. The stock solution is diluted with a diluent (sometimes also called a solvent), which may be water or some other suitable substance. Whenever a dilution is required, the necessary math may be performed up to three different ways, depending on what information is present, including a series of ratio proportions, the dilution formula, or the alligation method. The following example will be solved all three ways.

Example: If 500 milliliters of a 5% stock solution is ordered, how many milliliters of a 25% stock solution are needed to prepare the desired solution and how many milliliters of a suitable diluent are required?

Series of ratio proportions

Solving this with a series of ratio proportions requires the least explanation as the concept has already been explained earlier in this chapter. To solve the problem this way, you would first need to determine how much drug is actually needed for the desired solution, and then figure out how many milliliters of stock solution will be needed to provide the necessary quantity of drug. Lastly, you can subtract the milliliters of stock solution required from the final volume to determine the quantity of diluent needed.

$$\frac{5\,g}{100\,mL} = \frac{N}{500\,mL}$$

$$N = 25\,g\,of\,drug$$

$$\frac{25\,g}{100\,mL} = \frac{25\,g}{N}$$

$$N = 100\,mL\,of\,stock\,solution$$

$$500\,mL - 100\,mL = 400\,mL\,of\,diluent$$

Solving this with the dilution formula requires you to plug information into the following formula and solve for the variable. Once you know both the final volume and the stock solution volume, you can subtract the milliliters of stock solution required from the final volume to determine the quantity of diluent needed.

Dilution formula

$(C_1)(Q_1)=(C_2)(Q_2)$

C1 = concentration of available stock solution (percentage strength is a type of concentration)
Q1 = quantity of available stock solution needed
C2 = concentration of ordered solution
Q2 = quantity of ordered solution needed

So, if we plug the numbers from our example problem into this formula and solve it, it looks like this:

$$25\% \times Q_1 = 5\% \times 500\ mL$$

$$Q_1 = \textbf{100 mL of stock solution}$$

$$500\ mL - 100\ mL = \textbf{400 mL diluent}$$

Alligation method

The alligation method provides another way to solve this problem and may also be useful when you are mixing two different stock solutions together in order to get a third concentration. The following is an image with a brief description of how to set a problem up this way.

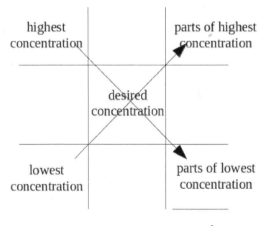

total parts

- Place the highest concentration in the upper left-hand corner.
- Place the lowest concentration in the lower left-hand corner (this will often be a diluent which will have a 0% concentration).
- Place the desired concentration in the center.
- Find the difference between the highest concentration and the desired concentration to find the parts of lowest concentration.

160

- Find the difference between the lowest concentration and the desired concentration to find the parts of highest concentration.
- Add the parts of highest concentration and the parts of lowest concentration to find the total parts.
- This provides you with a ratio that you can use to finish solving the problem.

With those concepts in mind, let's plug numbers in from our example question.

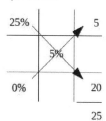

$$\frac{5}{25} \times 500 \ mL = \mathbf{100 \ mL \ of \ 25\% \ solution}$$

$$\frac{20}{25} \times 500 \ mL = \mathbf{400 \ mL \ of \ diluent}$$

Mixing liquid preparations

Sometimes, you will need to determine how much to use of various products to fulfill a recipe written by a physician; sometimes, you may take a recipe for a liquid medication and modify it for a different final volume; and other times you may need to either open up capsules or crush tablets, and dissolve or suspend them in a liquid vehicle. Let's look at an example of each scenario.

Example: How much clindamycin phosphate (the stock vial concentration is 150 mg/mL) and how much Cetaphil Lotion are needed to make the following compound?

Rx clindamycin phosphate 1200 mg in Cetaphil Lotion
Disp: 120 mL
Sig: aa hs ud

First, we should figure out how many mL of the clindamycin phosphate stock solution we should use:

$$\frac{1200 \ mg}{1} \times \frac{mL}{150 \ mg} = \mathbf{8 \ mL \ of \ clindamycin \ phosphate}$$

Then, we should figure out how much Cetaphil Lotion we'll need to make this 120 mL:

$$120\,mL - 8\,mL = 112\,mL\ of\ Cetaphil\ Lotion$$

Example: A prescription is written for a mouthwash containing 170 mL diphenhydramine elixir, 50 mL lidocaine viscous, 200 mL nystatin suspension, 52 mL erythromycin ethyl succinate suspension, and 28 mL of cherry syrup to make 500 mL of mouthwash. How much of each ingredient would be needed if you only wanted to prepare 240 mL of the mouthwash?

First, make a ratio comparing each ingredient and specifying the total volume:

170:50:200:52:28 to make 500 mL mouthwash

Then, solve for how much of each ingredient is needed to make 240 mL of mouthwash:

$$\frac{170\,mL\ diphenhydramine\ elixir}{500\,mL\ mouthwash} = \frac{N}{240\,mL\ mouthwash}$$
$$N = \textbf{81.6}\,mL\ diphenhydramine\ elixir$$

$$\frac{50\,mL\ lidocaine\ viscous}{500\,mL\ mouthwash} = \frac{N}{240\,mL\ mouthwash}$$
$$N = \textbf{24}\,mL\ lidocaine\ viscous$$

$$\frac{200\,mL\ nystatin\ suspension}{500\,mL\ mouthwash} = \frac{N}{240\,mL\ mouthwash}$$
$$N = \textbf{96}\,mL\ nystatin\ suspension$$

$$\frac{52\,mL\ erythromycin\ ethyl\ succinate\ suspension}{500\,mL\ mouthwash} = \frac{N}{240\,mL\ mouthwash}$$
$$N = \textbf{25}\,mL\ erythromycin\ ethyl\ succinate\ suspension$$

$$\frac{28\,mL\ cherry\ syrup}{500\,mL\ mouthwash} = \frac{N}{240\,mL\ mouthwash}$$
$$N = \textbf{13.4}\,mL\ cherry\ syrup$$

Example: How many 25 mg tablets of metoprolol tartrate and how many milliliters of Ora-Plus and Ora Sweet are needed to compound the following prescription?

Rx metoprolol tartrate 6.25 mg/tsp in a 50:50 mixture of Ora-Plus and Ora-Sweet

Disp: 300 mL
Sig: i tsp po bid

First, determine how many metoprolol tartrate tablets are needed:

$$\frac{tablet}{25\,mg}\times\frac{6.25\,mg}{tsp}\times\frac{tsp}{5\,mL}\times\frac{300\,mL}{1}=15\,tablets$$

You will often expect the powder volume from crushed tablets and opened capsules to be negligible, but since we don't know exactly, we will simply do the math for both liquids as if all the volume was from our two suspending agents. Since they are a 50:50 mixture, it means that we only need half the volume for each suspension.

$$\frac{50}{100}=\frac{N}{300\,mL}$$
$$N=150\,mL\,Ora-Plus$$

$$\frac{50}{100}=\frac{N}{300\,mL}$$
$$N=150\,mL\,Ora-Sweet$$

Compounding ointments gels and creams

Sometimes, compounding a semi-solid mixture (ointment, gel, or cream) can be as straight forward as mixing two semi-solids together. Other times, it may require incorporating a medication into a semi-solid base. Let's look at an example of each.

Example: A prescription is written for equal parts triamcinolone 0.1% cream and Lamisil cream, dispense 30 grams. How many grams of triamcinolone 0.1% cream are needed to fill the prescription? How many grams of Lamisil cream are needed to fill the prescription? What is the final percentage strength of triamcinolone in the compound?

To solve this, we need to first recognize that the ratio between the ingredients is 1:1, for a total of 2 parts. With that in mind, we know that half the total weight is how many grams of each ingredient we'll need.

$$\frac{1}{2}=\frac{N}{30\,g}$$
$$N=15\,g$$

Therefore, we will need **15 g of triamcinolone 0.1% cm** and **15 g of Lamisil cm**.

Next, we need to evaluate the final percentage strength of triamcinolone in the compound. There are 2 ways to do it: one is to calculate just how much triamcinolone is in the mixture, and then figure out its percentage strength; the other is to also divide by 2 like we did the total weight. Both ways will be demonstrated, but recognize that you only have to do it one way to achieve the correct answer.

$$\frac{0.1\,g}{100\,g}=\frac{N}{15\,g}$$
$$N=0.015\,g$$

$$\frac{0.015\,g}{30\,g}=\frac{N}{100\,g}$$
$$N=\textbf{0.05\% } \textit{triamcinolone}$$

or

0.1% ÷ 2 = **0.05% triamcinolone**

Obviously, the second way was easier, but it is good to know that you will get the same answer either way.

It is also noteworthy that the methodology used in this example will apply to any compounding problem where you are mixing ingredients in equal parts.

Example: If 50 g of salicylic acid ointment contains 10 grams of salicylic acid, what is the percentage strength of salicylic acid in the ointment?

This problem is just a simple w/w percentage strength problem:

$$\frac{10\,g\, salicylic\, acid}{50\,g\, ointment}=\frac{N}{100\,g\, ointment}$$
$$N=\textbf{20\% } \textit{salicylic acid}$$

Reconstituting powders

Some medications come from the manufacturer as a lyophilized powder in order to provide them with a longer shelf life. Some medications have a negligible powder volume, and others have a more significant volume. Whenever looking at a powder that has been reconstituted, you must be aware of the resulting concentration of the medication, whether it is an oral medication labeled as 400 milligrams per teaspoon, or injectable medication with a concentration of 250,000 units per milliliter. Let's look at an example with respect to a powder intended for reconstitution.

Example: If a 20 gram vial of cefazolin Na is reconstituted with 192 milliliters of sterile water and contains 8 milliliters of powder volume, how many milliliters of the reconstituted solution will be needed to provide a 1 gram dose?

$$192 \, mL \; diluent$$
$$+ \quad 8 \, mL \; powder \; volume$$
$$200 \, mL \; total \; volume$$

$$\frac{20 \, g}{200 \, mL} = \frac{1 \, g}{N}$$

$$N = 10 \, mL$$

Dosages based on body weight

Many drugs need to be calculated based on body weight. Some of the drugs where you will see this most often include chemotherapy, steroids, antibiotics, heparinoids, and drugs for pediatric and geriatric patients. If a medication is to be based on body weight, it will usually be requested in mg/kg. Look at the following example:

Example: A physician orders cyclophosphamide to be given 5 mg/kg in a 50 mL D5W bag. If the patient weighs 132 lbs and the concentration of the drug after reconstitution is 500 mg/10 mL, how many milliliters will you need to withdraw from the vial and add to the bag in order to prepare this admixture?

$$\frac{132 \, lbs}{1} \times \frac{1 \, kg}{2.2 \, lbs} \times \frac{5 \, mg}{kg} \times \frac{10 \, mL}{500 \, mg} = 6 \, mL$$

Dosages based on body surface area

Body surface area (BSA) is the measured or calculated surface of the human body, and it is measured in square meters (m^2). For many clinical purposes, BSA is a better indicator of metabolic mass than body weight because it is less affected by abnormal adipose mass. There are a number of weighs to calculate, but one of the simplest is the Mosteller formula.

Mosteller

$$BSA(m^2) = \sqrt{\frac{\text{weight in kg} \times \text{height in cm}}{3600}}$$

Example: If a patient is 6' 1" tall and weighs 176 lbs, what is the patient's BSA?

$$6\,\text{ft}\,1\,\text{in} = 73\,\text{in}$$

$$\frac{73\,\text{in}}{1} \times \frac{2.54\,\text{cm}}{1\,\text{in}} = 185.42\,\text{cm}$$

$$\frac{176\,\text{lbs}}{1} \times \frac{1\,\text{kg}}{2.2\,\text{lbs}} = 80\,\text{kg}$$

$$\sqrt{\frac{80\,\text{kg} \times 185.42\,\text{cm}}{3600}} = \mathbf{2.03}\ \boldsymbol{m^2}$$

Example: A physician orders bleomycin in a dose of 20 units/m^2 for a 5'3" patient that weighs 110 lbs. How many units of bleomycin should the patient receive?

$$5\,\text{ft}\,3\,\text{in} = 63\,\text{in}$$

$$\frac{63\,\text{in}}{1} \times \frac{2.54\,\text{cm}}{1\,\text{in}} = 160\,\text{cm}$$

$$\frac{110\,\text{lbs}}{1} \times \frac{1\,\text{kg}}{2.2\,\text{lbs}} = 50\,\text{kg}$$

$$\sqrt{\frac{50\,\text{kg} \times 160\,\text{cm}}{3600}} = 1.49\,m^2$$

$$\frac{1.49\,m^2}{1} \times \frac{20\,\text{units}}{m^2} = \mathbf{29.8\,units}$$

Quick review question 22

If your pharmacy stocks 20 milliliter vials of 2% lidocaine, how many milliliters would you need to draw up to fulfill a request for 100 mg of lidocaine?

a. 5 mL
b. 2 mL
c. 4 mL
d. 20 mL

Quick review question 23

How many milliliters of a 2% stock solution should be used to prepare 500 milliliters of a 0.025% solution?

a. 10 mL
b. 12.5 mL
c. 6.25 mL
d. none of the above

Quick review question 24

If a pharmacy compounds and dispenses 60 grams of a 2.5% hydrocortisone cream made from hydrocortisone powder and Eucerin cream, how much hydrocortisone should have been used to compound it?

a. 1.5 g
b. 2.5 g

c. 58.5 g
d. 60 g

Quick review question 25

The directions for a vial containing 1 gram of lyophilized ceftriaxone recommends reconstitution with 3.6 milliliters of diluent. If the vial has 0.4 milliliters of powder volume, what will be the resulting concentration in mg/mL after adding the suggested quantity of diluent?

a. 0.25 g/mL
b. 250 mg/mL
c. 278 mg/mL
d. 2500 mg/mL

Quick review question 26

A patient weighing 250 lbs is ordered infliximab 5 mg/kg. How many milligrams of infliximab should the patient receive?

a. 1250 mg
b. 568 mg
c. 113 mg
d. 50 mg

Quick review question 27

A 240 pound patient is admitted to the hospital due to herpes genitalis. The physician orders acyclovir 5 mg/kg in 100 mL of NS q8h. Acyclovir is available in 1 gram vials that get reconstituted with 20 milliliters of sterile water (powder volume is negligible). How many milliliters of the reconstituted acyclovir will you add to each 100 mL bag?

a. 545 mg
b. 1200 mg
c. 24 mL
d. 10.9 mL

Quick review question 28

An oncologist is initiating therapy with rituximab 375 mg/m^2 on a patient with chronic lymphocytic leukemia. The patient is 68" tall and weighs 140 lbs. How many milligrams of rituximab should the patient receive?

   a. 1.63 m^2
   b. 610 mg
   c. 1.75 m^2
   d. 656 mg

Answers to quick review questions

1. A - Chapters in the USP that are listed as below 1000 are considered enforceable, while chapters enumerated as 1000 or greater are considered guidelines.
2. B - USP Chapter 797, Pharmaceutical Compounding-Sterile Preparations, provides the first set of enforceable sterile compounding standards issued by the United States Pharmacopeia (USP).
3. C - A nosocomial infection is an infection acquired while in the hospital.
4. D - There are two types of pharmaceutical compounding: sterile (ophthalmics, IVs, parenteral nutrition, chemotherapy, and various other injectables) and nonsterile (creams, ointments, oral suspensions, capsules, suppositories, medication sticks, troches, etc.).
5. B - USP 797 provides requirements for how personnel should wash and garb, including the order that they should don their PPE.
6. D - Waste that poses no ecological threats, is devoid of protected health information, and poses no major safety risks may be disposed of in regular waste receptacles.
7. A - Broken ampules, syringes, and various needles need to be treated with caution to avoid accidental needle sticks and/or cuts. If the medications or diluents that these sharps were exposed to are not classified as hazardous, then you may dispose of this waste in a red sharps container.
8. B - Fully used vials, syringes, tubing, and bags of hazardous drug waste, along with PPE used while working with hazardous drugs, may be disposed in yellow, properly labeled containers. Also, any partially used or expired hazardous drugs that are not also considered to be RCRA

(Resource Conservation and Recovery Act) regulated hazardous drugs should be placed in these yellow buckets. Trace contaminated items, such as booties, gowns, and masks, even if they were not involved in spills, MUST be treated as hazardous waste (the yellow buckets are sufficient for this).

9. D - Whenever preparing any extemporaneous compound, the minimum information to maintain would be a method for tracking whom the compounder(s) is/are, who verified that the procedures for compounding were carried out correctly, and drugs/chemical used including manufacturer, lot number, expiration date, and an internally assigned batch number and beyond-use date.

10. B - A batch preparation is one in which multiple identical units are prepared in a single operation in anticipation of prescriptions.

11. D - Some incompatibilities are visible (color change or formation of a precipitate), and other chemical incompatibilities can be equally concerning even if not visible, such as the formation of gases and volatile chemicals.

12. B - The beyond-use date placed on the label shall be no later than the expiration date on the manufacturer's container. The beyond-use date is a date after which an article [drug] must not be used. Based on the information supplied by the manufacturer, the dispenser shall place on the label of the prescription container a suitable beyond-use date to limit the patient's use of the article.

13. C - Class III balances have a minimal weighable quantity (MWQ) of 120 mg.

14. C - When the balance has determined the final weight, press the tare bar to compensate for the weight of the weighing boat or weighing paper.

15. C - The general rule is to measure volumes no less than 20% of the capacity of a graduate.

16. D - Compounding pharmacies are capable of preparing a broad array of dosage forms including capsules, tablets, oral suspensions and solutions, suppositories, troches (lozenges), various topical dosage forms (gels, lotions, ointments, creams), lollipops, and more.

17. B - Medium-risk compounding is performed within an ISO Class 5 (or better) environment, and involves more complex manipulations of sterile products including more product manipulations and potentially more medication port entries than Low-risk compounding. As long as all products and equipment used directly in compounding it are sterile, it is still considered medium-risk.

18. A - A facility that prepares more than a low volume of hazardous drugs is required to have a separate negative pressure room in which an appropriate primary engineering control will be housed.

19. C - Examples of primary engineering control may include laminar airflow workbenches (LAFW), compounding aseptic isolators (CAI), biological safety cabinets (BSC), compounding aseptic containment isolators (CACI), and clean rooms that create an ISO Class 5 environment. A BCS is not an actual item used in compounding.

20. C - It is important to use a syringe that comes closest to the volume that needs to be measured, which ideally can also contain the full quantity desired.

21. C - High-risk compounding is performed within an ISO Class 5 (or better) environment, but involves the use of either nonsterile products and/or equipment. These medications must be terminally sterilized prior to use for a patient.

22. A - The problem can be solved as follows:

$$\frac{100\,\text{mg}}{1} \times \frac{1\,\text{g}}{1000\,\text{mg}} = 0.1\,\text{g}$$

$$\frac{0.1\,\text{g}}{N} = \frac{2\,\text{g}}{100\,\text{mL}}$$

$$N = 5\,\text{mL}$$

23. C - The problem can be solved as follows:

$$C_1 \times Q_1 = C_2 \times Q_2$$

$$2\% \times Q_1 = 0.025\% \times 500\,\text{mL}$$

$$Q_1 = 6.25\,\text{mL}$$

24. A - The problem can be solved as follows:

$$\frac{N}{60\,g} = \frac{2.5\,g}{100\,g}$$

$N = $ **1.5 g hydrocortisone**

25. B - The problem can be solved as follows:

    3.6 mL diluent
   + 0.4 mL powder volume
    4.0 mL total volume

$$\frac{1\,g}{4\,mL} \times \frac{1000\,mg}{1\,g} = \textbf{250 mg/mL}$$

26. B - The problem can be solved as follows:

$$\frac{250\,lbs}{1} \times \frac{1\,kg}{2.2\,lbs} \times \frac{5\,mg}{1\,kg} = \textbf{568 mg}$$

27. D - The problem can be solved as follows:

$$\frac{240\,lbs}{1} \times \frac{1\,kg}{2.2\,lbs} \times \frac{5\,mg}{1\,kg} \times \frac{1000\,mg}{1\,g} \times \frac{20\,mL}{1\,g} = \textbf{10.9 mL}$$

28. D - The problem can be solved as follows:

$$\frac{140\,lbs}{1} \times \frac{1\,kg}{2.2\,lbs} = 64\,kg$$

$$\frac{68\,in}{1} \times \frac{2.54\,cm}{1\,in} = 173\,cm$$

$$\sqrt{\frac{64 \times 173}{3600}} = 1.75\,m^2$$

$$\frac{1.75\,m^2}{1} \times \frac{375\,mg}{m^2} = \textbf{656 mg}$$

CHAPTER 4 Medication Safety

Key concepts

This chapter will cover the following knowledge areas to prepare you for the *Pharmacy Technician Certification Exam*:

- Error prevention strategies for data entry (e.g., prescription or medication order to correct patient)
- Patient package insert and medication guide requirements (e.g., special directions and precautions)
- Identify issues that require pharmacist intervention (e.g., DUR, ADE, OTC recommendation, therapeutic substitution, misuse, missed dose)
- Look-alike/sound-alike medications
- High-alert/risk medications
- Common safety strategies (e.g., tall man lettering, separating inventory, leading and trailing zeros, limit use of error prone abbreviations)

Terminology

To get started in this chapter, there are some terms that should be defined.

medication errors - A medication error is any incorrect or wrongful administration of a medication, such as a mistake in dosage or route of administration, failure to prescribe or administer the correct drug or formulation for a particular disease or condition, use of outdated drugs, failure to observe the correct time for administration of the drug, or lack of awareness of adverse effects of certain drug combinations.

mondegreen - A mondegreen is the mishearing or misinterpretation of a phrase as a result of near-homophony, in a way that gives it an unintended meaning.

risk evaluation and mitigation strategy - A risk evaluation and mitigation strategy (REMS) is a plan to manage a known or potential serious risk associated with a drug or biological product.

prospective drug utilization review - Prospective drug utilization review (ProDUR) require state Medicaid provider pharmacists to review Medicaid recipients' entire drug profile before filling their prescription(s). The ProDUR is intended to detect potential drug therapy problems.

MedWatch - MedWatch is the Food and Drug Administration's reporting system for an adverse event.

adverse drug event - An adverse drug event (ADE) is any undesirable experience temporally associated with the use of a medication.

Vaccine Adverse Event Reporting System - Vaccine Adverse Event Reporting System (VAERS) is a post-marketing safety surveillance program which collects information about adverse events (possible side effects) that occur after administration of vaccines.

Quick review question 1

Fill in the blank: An adverse drug event is an undesirable experience that takes place _____ with the use of a medication.

   a. temporarily
   b. close in time
   c. immediately
   d. none of the above

Quick review question 2

Which of the following is not an example of a medication error?

   a. giving a patient a medication intended for someone else
   b. giving a medication intended for PO use IV instead
   c. giving a patient an HCTZ 12.5 mg tablet instead of an HCTZ 12.5 mg capsule
   d. All of the above examples should be considered medication errors.

REMS provide a plan to manage a known or potential serious risk associated with a drug or biological product. What does the acronym 'REMS' stand for?

a. risk evaluation and mitigation strategy
b. rapid eye movement sleep
c. an American rock band from Athens, Georgia
d. a 76 km long river in Germany

Medication errors

A medication error is any incorrect or wrongful administration of a medication, such as a mistake in dosage or route of administration, failure to prescribe or administer the correct drug or formulation for a particular disease or condition, use of outdated drugs, failure to observe the correct time for administration of the drug, or lack of awareness of adverse effects of certain drug combinations. Causes of medication error may include difficulty in reading handwritten orders, confusion about different drugs with similar names, and lack of information about a patient's drug allergies or sensitivities.

The National Coordinating Council for Medication Error Reporting and Prevention (NCC MERP) has organized medication errors into four major groupings encompassing a total of nine categories (categories A through I):

- **No Error**
- *Category A:* Circumstances or events that have the capacity to cause error.
- **Error, No Harm**
  Category B: An error occurred, but the error did not reach the patient.
  Category C: An error occurred that reached the patient, but did not cause patient harm.
  Category D: An error occurred that reached the patient, and required monitoring to confirm that it resulted in no harm to the patient, and/or required intervention to preclude harm.
- **Error, Harm**

Category E: An error occurred that may have contributed to or resulted in temporary harm to the patient and required intervention.

Category F: An error occurred that may have contributed to or resulted in temporary harm to the patient, and required initial or prolonged hospitalization.

Category G: An error occurred that may have contributed to or resulted in permanent patient harm.

Category H: An error occurred that required intervention necessary to sustain life.

- **Error, Death**

  Category I: An error occurred that may have contributed to or resulted in the patient's death.

The following are some sobering facts related to medication errors:

- The Institute of Medicine (IOM) estimates that up to 98,000 people die annually in the United States from medical errors. (This does not include those seriously injured.)
- To go further, a study conducted by The Leapfrog Group estimates that 7,000 of these deaths are due to medication errors.
- According to Franklin et al, 1.5 million people in the United States are sickened, injured, or killed each year due to prescription errors.

Quick review question 4

Which NCC MERP classification does the following situation belong in? A physician orders an IV infusion of promethazine for a patient. The pharmacy sends just the ampule instead of mixing it in a minibag, and when the drug is administered, the nurse accidentally gives the medication intra-arterially causing the patient to lose their left arm from severe tissue death.

a. no error
b. error, no harm
c. error, harm
d. error, death

Quick review question 5

Which NCC MERP classification does the following situation belong in? A pharmacy technician notices that the generic captopril 12.5 mg tablets, and the carvedilol 12.5 mg tablets on the shelf are very similar in appearance and are located near each other.

a. no error
b. error, no harm
c. error, harm
d. error, death

Quick review question 6

Which NCC MERP classification does the following situation belong in? A pharmacy technician accidentally fills a prescription for hydrochlorothiazide 12.5 mg tablets with hydrochlorothiazide 12.5 mg capsules, but the checking pharmacist corrects it before the medication leaves the pharmacy.

a. no error
b. error, no harm
c. error, harm
d. error, death

Quick review question 7

Which NCC MERP classification does the following situation belong in? A pharmacy did not request allergy information from a new patient and filled a prescription for amoxicillin. The patient had a severe reaction to the penicillin derivative, went into anaphylactic shock, and died after their first dose at home.

a. no error
b. error, no harm
c. error, harm
d. error, death

Common error sources

According to a study of prescription errors conducted by MEDMARX and released in 2005, the following is a quick breakdown about the causes of these errors:

- Performance deficit (failure to act in accordance with

education and training)- 38%
- Procedure/protocol not followed - 17%
- Transcription inaccurate/omitted - 13%
- Computer entry - 12%
- Documentation lacking/inaccurate - 12%
- Knowledge deficit - 10%
- Communication - 9%

Error prone abbreviations

There are a number of abbreviations and dose designations that can be misinterpreted. Below is a list of some of the more commonly misinterpreted abbreviations and dose designations.

μg - microgram - mistaken as "mg" - to avoid confusion, use "mcg"
U or **u** - unit - mistaken for "0" (zero), the number "4" (four) or "cc" - to avoid confusion, write "unit"
IU - International Unit - mistaken for "IV" (intravenous) or the number "10" (ten) - to avoid confusion, write "International Unit"
Q.D., **QD**, **q.d.**, or **qd** - daily - mistaken for qod (every other day) or "qid" (four times a day) - to avoid confusion, write "Daily"
Q.O.D., **QOD**, **q.o.d.**, or **qod** - every other day - the period after the Q mistaken for "I" and the "O" mistaken for "I" causing confusion with "qid" - to avoid confusion, write "every other day"
q.i.d., or **qid** - four times a day - the "I" may be missed or mistaken for a "." or a "O" causing confusion with "qd" and "qod" - to avoid confusion, write "four times a day"
TIW or **tiw** - three times a week - misinterpreted as twice a week - to avoid confusion, write "three times a week" or list the specific days
Trailing zero (X.0 mg) - means X mg - the decimal point may be missed and read as "X0 mg" - to avoid confusion, don't use a trailing zero
Lack of leading zero (.X mg) - means 0.X mg - Without a leading zero, the decimal can be missed causing "X mg" to be dispensed - to avoid confusion, write "0.X mg"
MS - Can mean morphine sulfate or magnesium sulfate, and therefore can be misinterpreted - to avoid confusion, write "morphine sulfate" or "magnesium sulfate"
MSO4 and **MgSO4** - morphine sulfate and magnesium sulfate - they

can be confused for one another - to avoid confusion, write "morphine sulfate" or "magnesium sulfate"

Look-alike/sound-alike medications

A number of medications have names with similar spellings (hydroxyzine and hydralazine), similar appearance (captopril and carvedilol are both available as small round white tablets), and some medication names can be misheard for each other (Doribax and Zovirax are nearly homophones). A modern term for this last category is a mondegreen. Barcoding technologies and tallman lettering are two of the more common strategies used to overcome these challenges, and are discussed further later in this chapter.

High-alert/risk medications

High-alert medications are drugs that bear a heightened risk of causing significant patient harm when they are used in error. Although mistakes may or may not be more common with these drugs, the consequences of an error are clearly more devastating to patients. The Institute for Safe Medication Practice (ISMP) has assembled a list to identify which medications require special safeguards to reduce the risk of errors. This may include strategies such as standardizing the ordering, storage, preparation, and administration of these products; improving access to information about these drugs; limiting access to high-alert medications; using auxiliary labels and automated alerts; and employing redundancies, such as automated or independent double-checks when necessary. (Note: manual independent double-checks are not always the optimal error-reduction strategy, and may not be practical for all of the medications on the list).

Classes/categories of high-alert medications

The following list of classes/categories of high-alert medications come form the ISMP:

- adrenergic agonists, IV (e.g., epinephrine, phenylephrine, norepinephrine)
- adrenergic antagonists, IV (e.g., propranolol, metoprolol,

labetalol)
- anesthetic agents, general, inhaled and IV (e.g., propofol, ketamine)
- antiarrhythmics, IV (e.g., lidocaine, amiodarone)
- antithrombotic agents, including:
  - anticoagulants (e.g., warfarin, low-molecular-weight heparin, IV unfractionated heparin)
  - Factor Xa inhibitors (e.g., fondaparinux)
  - direct thrombin inhibitors (e.g., argatroban, bivalirudin, dabigatran etexilate, lepirudin)
  - thrombolytics (e.g., alteplase, reteplase, tenecteplase)
  - glycoprotein IIb/IIIa inhibitors (e.g., eptifibatide)
- cardioplegic solutions
- chemotherapeutic agents, parenteral and oral
- dextrose, hypertonic, 20% or greater
- dialysis solutions, peritoneal and hemodialysis
- epidural or intrathecal medications
- hypoglycemics, oral
- inotropic medications, IV (e.g., digoxin, milrinone)
- insulin, subcutaneous and IV
- liposomal forms of drugs (e.g., liposomal amphotericin B) and conventional counterparts (e.g., amphotericin B desoxycholate)
- moderate sedation agents, IV (e.g., dexmedetomidine, midazolam)
- moderate sedation agents, oral, for children (e.g., chloral hydrate)
- narcotics/opioids including:
  - IV
  - transdermal
  - oral (including liquid concentrates, immediate and sustained-release formulations)
- neuromuscular blocking agents (e.g., succinylcholine, rocuronium, vecuronium)
- parenteral nutrition preparations
- radiocontrast agents, IV
- sterile water for injection, inhalation, and irrigation (excluding pour bottles) in containers of 100 mL or more

- sodium chloride for injection, hypertonic, greater than 0.9% concentration

Specific high-alert medications

The following list of specific high-alert medications come form the ISMP.

- epoprostenol (Flolan), IV
- magnesium sulfate injection
- methotrexate, oral, non-oncologic use
- opium tincture
- oxytocin, IV
- nitroprusside sodium for injection
- potassium chloride for injection concentrate
- potassium phosphates injection
- promethazine, IV
- vasopressin, IV or intraosseous

Quick review question 8

Which of the following is the best practice for requesting the following levothyroxine prescription?

a. levothyroxine 25 µg tab, Disp 30, Sig: i tab po daily
b. levothyroxine .025 mg tab, Disp 30, Sig: 1 tab po daily
c. levothyroxine 25 mcg tab, Disp 30, Sig: 1 tab po qd
d. None of the examples follow best practices.

Quick review question 9

Why might Zovirax and Doribax be mistaken for each other?

a. They have similar therapeutic indications.
b. They have near-homophony names.
c. They have similar appearance.
d. No one would ever mistake them for each other.

Quick review question 10

According to the ISMP, what are high-alert medications?

a. They are drugs that bear a heightened risk of causing significant patient harm when they are used in error.
b. They only consist of medications with published REMS.

c. They are medications that should be placed on the top shelf to emphasize their dangerous nature.

d. all of the above

Quick review question 11

When considering sources of medication errors, what is a performance deficit?

a. package failure causing product adulteration

b. underdosing of medication (whether by patient or physician) resulting in a poor therapeutic outcome

c. failure to act in accordance with education and training

d. all of the above

Quick review question 12

Which of the following is a high-alert medication?

a. acyclovir
b. penicillin
c. upsidaisium
d. warfarin

Quick review question 13

Which of the following classes is not a high-alert medication?

a. chemotherapeutic agents
b. epidural or intrathecal medications
c. injectable antibiotics
d. neuromuscular blocking agent

Quick review question 14

A physician orders a continuous infusion of gentamicin for a patient, and due to the drugs ototoxicity, the patient ends up with a partial loss of hearing. What is the most likely source of the error?

a. transcription inaccurate/omitted
b. computer entry
c. documentation lacking/inaccurate
d. knowledge deficit

Quick review question 15

The abbreviation MS should be avoided because it can be interpreted as which of the following?

a. magnesium sulfate
b. morphine sulfate
c. both A and B
d. none of the above

Quick review question 16

What is a high-alert medication?

a. a drug that bears a heightened risk of causing significant patient harm when they are used in error
b. schedule I medications
c. recalled medications that have a high risk of causing death or permanent harm
d. none of the above

Error prevention strategies

The medical profession has a responsibility to provide safe and effective health care, and that includes reducing medication misadventures. There are a broad array of error prevention tools out there, including using the five rights (some times also expanded to the seven rights), using extra precautions with respect to high-alert medications, adopting system wide use of Tallman lettering, avoiding error prone abbreviations, using barcoding technology, and even just separating inventory. Providing information to patients through medications guides, patient counseling, and drug utilization reviews (DURs) improves compliance and decreases errors as well.

Five rights

There are five rights commonly associated with medication administration to help improve patient safety.

The five rights of medication administration are:

- *Right patient* - Most important of all, do you have the right patient? Verify the patient information. Check the wrist band.

Ask the patient to tell you his name and date of birth. If he can't tell you, does he know his physician's name? Is there a family member present to verify identification? Never use the room and bed number as the sole means of identification.

- *Right medication* - Check the label on the bottle, bubble pack or other packaging. Also be aware of the generic name of the medication ordered in case of product substitution.
- *Right dose* - Make sure that the quantity and strength of the medication matches what was ordered.
- *Right time* - Many medications are time sensitive. Be sure to give the medication at the proper time and at the proper interval.
- *Right route* - Some medications have various routes that they could be administered. Be sure that the medication you have is appropriate for the route it needs to be administered.

Many institutions have started including additional rights, which vary from facility to facility. These additional rights most typically include:

- *Right technique* - Be sure you are using the correct technique to administer the medication. Typically, something intended for intramuscular injection will not achieve the desired effect if given intravenously.
- *Right chart information* - In an institutional setting, there will usually be a patient chart that you can verify all your information against in order to double-check everything else.
- *Right documentation* - Proper documentation of information will help the other health professionals provide better care for the patient when they look over the patient's chart/medication administration record.
- *Right attitude* - Patients and their family members will typically be much more compliant and helpful if the medical care givers display an appropriate attitude.

Tallman lettering

Tallman lettering (or Tall Man lettering) is the practice of writing part of a drug's name in upper case letters to help distinguish sound-alike, look-alike drugs from one another in order to avoid medication errors. Several studies have shown that highlighting sections of drug names using tallman letters can help distinguish similar drug names, making them less prone to mix-ups. ISMP, FDA, The Joint Commission, and other safety-conscious organizations have promoted the use of tallman letters as one means of reducing confusion between similar drug names.

Below are two lists with recommendations for the use of tallman lettering. The first list is of FDA-approved established drug names with recommended tallman letters, which were first identified during the FDA Name Differentiation Project. The second list is of additional drug names with recommendations from ISMP regarding the use and placement of tallman letters.

FDA's list

The following is a Food and Drug Administration approved list of generic drug names with tallman letters. The medications have been grouped to emphasize which medications are commonly mistaken by having sound-alike, look-alike names.

acetaHEXAMIDE
acetaZOLAMIDE

buPROPion
busPIRone

chlorproMAZINE
chlorproPAMIDE

clomiPHENE
clomiPRAMINE

cycloSPORINE
cycloSERINE

DAUNOrubicin
DOXOrubicin

dimenhyDRINATE
diphenhydrAMINE

DOBUTamine
DOPamine

glipiZIDE
glyBURIDE

hydrALAZINE
HYDROmorphone
hydrOXYzine

medroxyPROGESTERone
methylPREDNISolone
methylTESTOSTERone
mitoXANTRONE

niCARdipine
NIFEdipine

predniSONE
prednisoLONE

risperiDONE
rOPINIRole

TOLAZamide
TOLBUTamide

sulfADIAZINE
sulfiSOXAZOLE

vinBLAStine
vinCRIStine

ISMP's list

The following is an Institute for Safe Medication Practices recommended list of drug names with tallman letters. The medications have been grouped to emphasize which medications are commonly mistaken by having sound-alike, look-alike names.

ALPRAZolam
LORazepam

CeleBREX
CeleXA

aMILoride
amLODIPine

chlodiazePOXIDE
chlorproMAZINE

ARIPiprazole
RABEprazole

clonazePAM
cloNIDine
cloZAPine
KlonoPIN
LORazepam

AVINza
INVanz

azaCITIDine
azaTHIOprine

DACTINomycin
DAPTOmycin

carBAMazepine
OXcarbazepine

DOCEtaxel
PACLitaxel

CARBOplatin
CISplatin

DOXOrubicin
IDArubicin

ceFAZolin
cefoTEtan
cefOXitin
cefTAZidime
cefTRIAXone

DULoxetine
FLUoxetine
PARoxetine

ePHEDrine
EPINEPHrine

fentaNYL
SUFentanil

flavoxATE
fluPHENAZine
fluvoxaMINE

guaiFENesin
guanFACINE

HomaLOG
HumuLIN

HYDROcodone
oxyCODONE
OxyCONTIN

HYDROmorphone
morphine

inFLIXimab
riTUXimab

ISOtretinoin
tretinoin

LaMICtal
LamISIL

lamiVUDine
lamoTRIgine

levETIRAcetam
levOCARNitine

metFORMIN
metroNIDAZOLE

mitoMYcin
mitoXANtrone

NexAVAR
NexIUM

niCARdipine
NIFEdipine
niMODipine

NovoLIN
NovoLOG

OLANZipine
QUEtiapine

PEMEtrexed
PRALAtrexate

PENTobarbital
PHENobarbital

PriLOSEC
PROzac

quiNIDine
quiNINE

RisperDAL
risperiDONE
rOPINIRole

romiDEPsin
romiPLOStim

SandIMMUNE
SandoSTATIN

SEROquel
SINEquan

sitaGLIPtin
SUMAtriptan
ZOLMitriptan

Solu-CORTEF
Solu-MEDROL

SORAfenib
SUNItinib

sulfADIAZINE
sulfaSALAzine

TEGretol
TRENtal

tiaGABine
tiZANidine

traMADol
traZODone

valACYclovir
valGANciclovir

ZyPREXA
ZyrTEC

Barcoding technologies

Barcoding technology, while common in many industries, was slower to emerge in health care (although in recent years, it has become much more common). Various studies have shown the potential for barcodes to reduce medication errors when dispensing medications from the pharmacy. Bedside barcode scanning, the use of barcode technology to verify a patient's identity and the medication to be administered, is a promising strategy for preventing medication errors. Also, many community pharmacies will scan the barcodes on the manufacturer's bottles to verify that the same NDC number entered in the pharmacy's computer system matches the NDC number on the drug being dispensed. Its use has been increasing in notable places such as the Veterans Affairs hospitals, University of Pittsburgh Medical Centers (UPMC), many of the major community pharmacy chains, and various hospitals throughout the United States.

Barcode medication verification at the bedside is usually implemented in conjunction with an electronic medication administration records (eMAR), allowing health care providers to automatically document the administration of drugs by means of barcode scanning. Because the eMAR imports medication orders electronically from either the computerized prescriber order entry (CPOE) or the pharmacy management software, its implementation may reduce transcription errors. Given its potential to improve medication safety, barcoding in combination with eMAR technology is being considered as a criterion for achieving "meaningful use" of health information technology, and for obtaining financial incentives under the American Recovery and Reinvestment Act of 2009 in 2013.

Risk evaluation and mitigation strategies

On September 27, 2007, President George W. Bush signed into law the Food and Drug Administration Amendments Act of 2007 (FDAAA), further amending the Food, Drug, and Cosmetic Act. The FDAAA authorizes the Food and Drug Administration (FDA) to require risk evaluation and mitigation strategies (REMS) on medications, if necessary, to minimize the risks associated with some drugs.

The FDA has a responsibility to ensure that the benefits of a medication outweigh the potential risks of that medication. To determine this, the FDA looks at the seriousness of the disease or condition to be treated, size of the patient population, expected benefit of the drug, expected duration of treatment, and seriousness of the known or potential adverse events.

If a medication's benefits outweigh the risks associated with it, the FDA approves it. Sometimes though, the medication therapy itself still carries a lot of risks with it. In that situation, the FDA may require REMS. These risk minimization strategies go beyond FDA professional labeling. FDA can require a REMS before a drug is approved if they have determined that a REMS is necessary to ensure the benefits of the drug outweigh its risks. Also, FDA can require a REMS post-approval if they become aware of new safety information, and make the determination that a REMS is necessary, again, to ensure that the benefits of the drug outweigh its risks.

REMS may contain any of the following elements:

- *Medication Guide* – Document written for patients highlighting important safety information about the drug; this document must be distributed by the pharmacist to every patient receiving the drug.

- *Communication Plan* – Plan to educate healthcare professionals on the safe and appropriate use of the drug, and consists of tools and materials that will be disseminated to the appropriate stakeholders.
- *Elements to Assure Safe Use (EASU)* – These are strictly controlled systems or requirements put into place to enforce the appropriate use of a drug. Examples of EASUs include physician certification requirements in order to prescribe the drug, patient enrollment in a central registry, distribution of the drug restricted to certain specialty pharmacies, etc.
- *Implementation Plan* – A description of how certain EASUs will be implemented.
- *Timetable for Submission of Assessments* – The frequency of assessment of the REMS performance with regard to meeting the goal(s) and objective(s). FDA requires that assessments be conducted at 18 months, 3 years, and 7 years post-launch, at a minimum. Results of these evaluations must be reported to the FDA, and will determine whether additional actions or modifications to the REMS program are required.

Patient package insert/medication guide requirements

A patient package insert or medication guide is a document provided along with a prescription medication to provide additional information about that drug. By providing this information to the patient, there is an expectation of improved therapeutic outcomes by improving compliance and helping the patient to avoid some potential errors from medication misuse.

The Food and Drug Administration (FDA) determines the requirements for patient package inserts. Package inserts follow a standard format for every medication and include the same types of information. Different manufacturers may have different titles for their sections, however, to make them easier for the average person to read and comprehend—for example, instead of "Contraindications" the section may be headed, "Who should not take this medication?"

The first thing listed is usually the brand name and generic name of the product. The other sections are as follows:

- Clinical pharmacology - tells how the medicine works in the body, how it is absorbed and eliminated, and what its effects are likely to be at various concentrations. May also contain results of various clinical trials (studies) and/or explanations of the medication's effect on various populations (e.g. children, women, etc.).
- Indications and usage - uses (indications) for which the drug has been FDA-approved (e.g. migraines, seizures, high blood pressure). Physicians legally can, and often do, prescribe medicines for purposes not listed in this section (so-called "off-label uses").
- Contraindications - lists situations in which the medication should not be used; for example, in patients with other medical conditions such as kidney problems or allergies.
- Warnings - covers possible serious side effects that may occur.
- Precautions - explains how to use the medication safely, including physical impairments and drug interactions; for example, "Do not drink alcohol while taking this medication", or "Do not take this medication if you are currently taking MAO inhibitors".
- Adverse reactions - lists all side effects observed in all studies of the drug (as opposed to just the dangerous side effects which are separately listed in "Warnings" section).
- Drug abuse and dependence - provides information regarding whether prolonged use of the medication can cause physical dependence (only included if applicable).
- Overdosage - gives the results of an overdose, and provides recommended action in such cases.
- Dosage and administration - gives recommended dosage(s); may list more than one for different conditions or different patients (e.g., lower dosages for children).
- How supplied - explains in detail the physical characteristics of the medication including color, shape, markings, etc., and storage information (e.g., "Do not store above 95° F").

Prospective drug utilization reviews

The Omnibus Budget Reconciliation Act of 1990 (OBRA-90) placed expectations on the pharmacist in how to interact with the patient. While the primary goal of OBRA-90 was to save the federal government money by improving therapeutic outcomes, the method to achieve these savings was implemented by imposing on the pharmacist counseling obligations, prospective drug utilization review (ProDUR) requirements, and record-keeping mandates.

These ProDURs require state Medicaid provider pharmacists to review Medicaid recipients' entire drug profile before filling their prescription(s). The ProDUR is intended to detect potential drug therapy problems. Computer programs can be used to assist the pharmacist in identifying potential problems. It is up to the pharmacist's professional judgment, however, as to what action to take, which could include contacting the prescriber. As part of the ProDUR, the following are areas for drug therapy problems that the pharmacist must screen:

- Therapeutic duplication
- Drug–disease contraindications
- Drug–drug interactions
- Incorrect drug dosage
- Incorrect duration of treatment
- Drug–allergy interactions
- Clinical abuse/misuse of medication

Patient counseling

When pharmacists assist patients with information on their medication use, they should include over the counter medications and dietary supplements as well. Technicians may assist in this process by identifying patients that could benefit from counseling and by gathering information, but all actual drug advice given to the patient should come from the pharmacist.

To take this a step further, the Medicare Modernization Act of 2003 allows for a pharmacist to provide (and get reimbursed for) medication therapy management (MTM), or an annual in-depth review of the Medicare patient's medication profile. This review is to add a safety feature to prevent adverse reactions and drug interactions, and to look at ways to reduce the patient cost.

Separating inventory

Separating inventory can be a useful tool for preventing errors by making harder to grab the wrong product, whether it is a concern over incorrect dosage form, or an error caused by sound-alike, look-alike drugs being near each other on a shelf. Pharmacies will separate hazardous drug storage from nonhazardous drugs, and internal products from external products. Many pharmacies will even take extra steps to separate sound-alike, look-alike drugs, whether by physically placing them on separate shelves, or through the use of automation to dispense (such as a medstation) to maintain them apart from each other.

Quick review question 17

In order to prevent medication errors, most institutions have implemented the '5 rights'. What are the '5 rights'?

a. right patient, right medication, right dose, right time, and right route
b. right DEA number, right NPI, right controlled substance schedule, right medication quantity, and right number of refills for state and federal guidelines
c. right name, right address, right birth date, and right social security number
d. right patient history, right allergy information, right diagnosis, right therapy and duration of treatment

Quick review question 18

What is tallman lettering?

a. printing a patient package insert in a larger font
b. write part of a drug's name in upper case letters
c. it is the process of writing men's names in upper case letters
d. none of the above

Quick review question 19

Which organization(s) promote the use of tallman letters to reduce medication errors?

a. FDA
b. ISMP
c. TJC
d. all of the above

Quick review question 20

The barcode on the side of manufacturer's medication bottle can be used to verify which of the following?

a. DEA number
b. lot number
c. NDC
d. ISBN

Quick review question 21

Placing clonidine and clozapine on different shelving units is an example of which kind of strategy to reduce medication errors?

a. tallman lettering
b. risk evaluation and mitigation strategy
c. barcoding technology
d. separating inventory

Quick review question 22

Labeling the shelves where clonidine and clozapine are kept as cloNIDine and cloZAPine is an example of which kind of strategy to reduce medication errors?

a. tallman lettering
b. risk evaluation and mitigation strategy
c. barcoding technology
d. separating inventory

Quick review question 23

Which organization(s) regulate risk evaluation and mitigation strategies?

a. FDA
b. ISMP

c. TJC
d. all of the above

Quick review question 24

A drug utilization review (DUR) should screen for which of the following items?

a. therapeutic duplication
b. drug–disease contraindications
c. drug–drug interactions
d. all of the above

Quick review question 25

A drug utilization review (DUR) should screen for which of the following items?

a. incorrect drug dosage
b. incorrect duration of treatment
c. drug–allergy interactions
d. all of the above

Quick review question 26

A drug utilization review (DUR) should screen for which of the following items?

a. clinical abuse/misuse of medication
b. affordability
c. third party reimbursement
d. all of the above

Quick review question 27

Tallman lettering is primarily used to distinguish between which of the following?

a. sound-alike, look-alike drugs
b. different strengths of the same medication
c. different dosage forms of the same medication
d. all of the above

Quick review question 28

Which organization determines the requirements of a patient

package insert?

a. manufacturer
b. FDA
c. ISMP
d. TJC

Quick review question 29

What is/are the intended purpose(s) of providing medication guides and patient package inserts?

a. reduction of medication errors
b. improved patient compliance
c. better therapeutic outcomes
d. all of the above

Quick review question 30

A patient with a history of atrial fibrillation has prescriptions for Tikosyn and warfarin, enters the pharmacy and asks "Would it be appropriate to use aspirin for a headache that started this morning?" Why is the pharmacist likely to tell the patient to use a different pain reliever?

a. drug-disease contraindications
b. over the counter aspirin would be the incorrect strength
c. drug-drug interactions
d. actually the patient should use aspirin, because it is more affordable than Tikosyn

Quick review question 31

If a customer comes to the pharmacy counter and asks for advice about over the counter medication, what should the technician do?

a. the technician should give advice
b. have the pharmacist counsel the patient
c. the patient should be told to contact their PCP with such questions
d. ask the patient if they have access to the internet

Quick review question 32

The in-depth reviews of a Medicare patient's medication profile is intended to do which of the following?

- a. prevent adverse reactions
- b. prevent drug interactions
- c. reduce the patient cost
- d. all of the above

Error reporting

The National Medication Errors Reporting Program, operated by the Institute for Safe Medication Practices (ISMP MERP), is a confidential voluntary program that provides expert analysis of system-based causes of medication errors. Through various media outlets, the Institute for Safe Medication Practices (ISMP) communicates recommendations for improving safety. ISMP MERP's voluntary reporting is available at https://www.ismp.org/orderforms/reporterrortoismp.asp.

It is also useful to report medication misadventures to the Food and Drug Administration's MedWatch program at http://www.fda.gov/Safety/MedWatch/default.htm, excluding vaccines. Adverse events involving vaccines should be reported to Vaccine Adverse Event Reporting System (VAERS) at http://vaers.hhs.gov/esub/index, which is run by both the FDA and the CDC. Much like ISMP MERP, MedWatch is also voluntary, but VAERS is actually mandatory.

MedWatch

MedWatch is the Food and Drug Administration's reporting system for an adverse event (sometime also called a sentinel event), and was founded in 1993. An adverse event is any undesirable experience temporally associated with the use of a medical product. The MedWatch system collects reports of adverse reactions and quality problems, primarily with drugs and medical devices, but also for other FDA-regulated products (e.g., dietary supplements, cosmetics, medical foods, and infant formulas). Vaccines are not covered by MedWatch, but instead entails the Vaccine Adverse Event Reporting System (VAERS). MedWatch also does not cover veterinary medicine and internet fraud.

FDA's MedWatch program offers voluntary reporting by healthcare professionals, consumers, and patients. This reporting can be conducted several different ways, including a single, one-page reporting form (Form FDA 3500) that can be mailed to the address on the form, or faxed to 1-800-FDA-0178; online reporting is available at https://www.accessdata.fda.gov/scripts/medwatch/medwatch-online. htm , or by phone at 1-800-FDA-1088.

The MedWatch system is intended to detect safety hazard signals for medical products. If a signal is detected, the FDA can issue medical product safety alerts or order product recalls, withdrawals, or labeling changes to protect the public health. Important safety information is disseminated to the medical community and the general public via the MedWatch web site at http://www.fda.gov/Safety/MedWatch/default.htm, and the MedWatch E-list and texting service that can be subscribed to at https://public.govdelivery.com/accounts/USFDA/subscriber/new? pop=t&topic_id=USFDA_46.

Raw data from the MedWatch system, together with adverse drug reaction reports from manufacturers as required by regulation, are part of a public database. Online tools that analyze the database are available for both health care consumers and professionals. The database is open enough that it was even used by journalists to investigate the FDA's drug approval practice.

History of MedWatch

Before David Kessler, former FDA Commissioner, introduced MedWatch in 1993, there was not a simple mechanism in place for healthcare professionals to report adverse events associated with medications or devices directly to the FDA. Many facilities and drug manufacturers had their own complicated forms, along with the FDA having separate forms for drug reactions, drug quality products problems, device quality product problems, and adverse reactions to medical devices creating a patchwork of reporting tools that were largely ignored. This was despite the post marketing surveillance required of all FDA approved medications.

Various studies conducted in the 1980's suggested that 3 to 11% of admissions could be attributed to adverse drug events, and yet only about 1% of adverse events were being reported to the FDA. This means that out of 1400 hospital admissions, approximately 100 would be related to adverse drug events, and yet only 1 was actually being reported.

MedWatch was intended to simplify this process, and the FDA was also trying to actively engage physicians, hospitals, and schools involved in health care education to use this new voluntary system. MedWatch encourages health care professionals to regard reporting as a fundamental professional and public health responsibility.

Why adverse event reporting is needed

Many health care professionals and their patients do not know that in approving medical products for sale and use (including medications, biologics and devices) that the FDA does not develop or test the products itself. FDA reviews the results of laboratory, animal and human clinical testing done by companies to determine if the product they want to put on the market is safe and effective.

Because the FDA's review is based on a relatively small number of users, and because variations in quality can happen in manufacturing, the FDA keeps careful watch on reports of adverse experiences with products after they are marketed. If this monitoring turns up a problem that needs to be corrected, the FDA can take timely actions to improve the safety of the product.

The FDA's role postmarketing surveillance when it is monitoring for safety is a critical one because of the limitations of pre-marketing review and the controlled clinical trials used to prove the effectiveness of the product before it is approved by the FDA.

For example, the reliance of the FDA and the manufacturer on the controlled clinical trial to demonstrate the efficacy of a drug for its intended use in a controlled population during NDA approval is of great use. It can be determined with some certainty that the product is effective when compared with placebo, and the common serious adverse reactions will be identified. But, limitations, such as number of participants and subject characteristics, results in uncertainties about the safety of the pharmaceutical once it is marketed and used in a wider population, over longer periods of time, in patients with co-morbidities and concomitant medications, and for 'off-label' uses not previously evaluated. This is a weakness of the clinical trial as a safety identification process. Therefore, the true picture of product safety actually evolves over the years that make up a product's lifetime in the marketplace.

How MedWatch works

The FDA is interested in receiving three types of reports. First, those involving suspected serious adverse events associated with drugs (either prescription or OTC), biologic products, medical devices, cosmetics and special nutritional products (such as dietary supplements, medical foods and infant formulas). Second, reports (especially from pharmacists and nurses) of product quality problems that might suggest manufacturing or other quality problems with drugs or devices. In addition, the FDA is very interested in reports of suspected counterfeit products. Third, reports of medication and device use errors (for example, the wrong drug/wrong dose errors that may be caused by either product name confusion or confusion resulting from packaging and labeling).

This reporting can be conducted several different ways, including a single, one-page reporting form (Form FDA 3500) that can be mailed to the address on the form or faxed to 1-800-FDA-0178; online reporting is available at https://www.accessdata.fda.gov/scripts/medwatch/medwatch-online.htm , or by phone at 1-800-FDA-1088. The four core elements of the report include: a reporter's name, a suspect drug or device product, a narrative report of the adverse event or problem, and an identifiable patient. The FDA, however, holds the identity of patients in strict confidence, protected by federal law and regulation.

Patients are encouraged to take this form to their health care professional, so that the report contains more detailed information when received by the FDA; however, patients may submit reports directly to the FDA.

Voluntary versus mandatory reporting

Often, there is a lot of concern over MedWatch's voluntary reporting, as opposed to implementing a mandatory adverse event reporting provision. The challenge would be enforcing such a provision. For a program to be mandatory, there would have to be a penalty involved with noncompliance. The FDA is concerned that penalizing individuals might deter participation rather than increase it.

An example of MedWatch's effectiveness

The story of the drug Depakote (divalproex sodium), illustrates that even after twenty years of use, new information that will improve the safe use of a medical product will be detected with the help of voluntary reports from health professionals.

In July 2000, the FDA added new safety information, a black boxed warning, for Depakote after receiving multiple reports of a serious adverse event, pancreatitis, in patients taking the drug. Although the condition was listed as a possible adverse reaction when first approved in 1978, the severity of this serious condition was found to be much greater when used more widely for newer and common indications such as migraines, depression, and seizures. It was with this increased use that spontaneous reports began to be received at the FDA through its MedWatch program, prompting the new warnings to inform doctors and patients of this risk.

Even after twenty years, new information that will impact the clinical use of a medical product can be detected. Consequently, all medical products need to be continually assessed for safety.

VAERS

The Vaccine Adverse Event Reporting System is a United States program for vaccine safety, co-managed by the Centers for Disease Control and Prevention (CDC) and the Food and Drug Administration (FDA). VAERS is a post-marketing safety surveillance program, collecting information about adverse events (possible side effects) that occur after administration of vaccines.

The National Childhood Vaccine Injury Act (NCVIA) of 1986 requires health professionals and vaccine manufacturers to report to the U.S. Department of Health and Human Services (HHS) specific adverse events that occur after the administration of routinely recommended vaccines. In response to NCVIA, CDC and FDA established VAERS in 1990.

VAERS has demonstrated its public health importance by providing health scientists with signals about possible adverse events following immunization. In one instance, VAERS detected reports for intussusception over that what would be expected to occur by chance alone after the RotaShield rotavirus vaccine in 1999. Epidemiologic studies confirmed an increased risk, and these data contributed to the product's removal from the US market. In another example, VAERS determined that there may be a potential for a small increase in risk for Guillain-Barre' syndrome (GBS) after the meningococcal conjugate vaccine, Menactra. As a result of this finding, a history of GBS became a contraindication to the vaccine, and further controlled studies are currently underway to research this issue.

Reporting

You are required to report all significant adverse events that occur after vaccination of adults and children, even if you are not sure whether the vaccine caused the adverse event. There are three ways to report to VAERS including online, by fax, and by mail. The online form can be found at https://vaers.hhs.gov/esub/step1 . You may also printout the VAERS form and fax it to (877) 721-0366 or mail it to VAERS, P.O. Box 1100, Rockville, MD 20849-1100. A pre-paid postage stamp is included on the back of the printed form.

Number of reports VAERS receives

VAERS receives around 30,000 reports annually, with 13% classified as serious (e.g., associated with disability, hospitalization, life-threatening illness or death) (CDC VAERS Master Search Tool, April 2, 2008). Since 1990, VAERS has received over 200,000 reports, most of which describe mild side effects such as fever. Very rarely, people experience serious adverse events following immunization. By monitoring such events, VAERS helps to identify any important new safety concerns, and thereby assists in ensuring that the benefits of vaccines continue to be far greater than the risks.

Many different types of adverse events occur after vaccination. About 85-90% of the reports describe mild adverse events such as fever, local reactions, and episodes of crying or mild irritability. The remaining reports reflect serious adverse events involving life-threatening conditions, hospitalization, permanent disability, or death, which may or may not have been caused by a vaccine.

Objectives of VAERS

The primary objectives of VAERS are to:

- Detect new, unusual, or rare vaccine adverse events (VAEs);
- Monitor increases in known adverse events;
- Identify potential patient risk factors for particular types of adverse events;
- Identify vaccine lots with increased numbers or types of reported adverse events; and
- Assess the safety of newly licensed vaccines.

Limitations

VAERS has several limitations, including unverified reports, underreporting, inconsistent data quality, and absence of an unvaccinated control group for comparisons to be made against.

Quick review question 33

Fill in the blank: Reporting medication adverse events into the MedWatch program is _____.

a. not recommended
b. voluntary
c. mandated
d. considered a misuse of the MedWatch program

Quick review question 34

The MedWatch program is regulated by which organizations?

a. CDC
b. FDA
c. both a and b
d. none of the above

Quick review question 35

The VAERS program is regulated by which organization(s)?

a. CDC
b. FDA
c. both a and b
d. none of the above

Quick review question 36

Fill in the blank: Reporting vaccine adverse events into the VAERS program is _____.

   a. not recommended
   b. voluntary
   c. mandated
   d. considered a misuse of the VAERS program

Quick review question 37

Which of the following programs are used in error reporting?

   a. MedWatch
   b. MERP
   c. VAERS
   d. all of the above

Quick review question 38

Whom may submit an error report to MedWatch?

   a. physicians
   b. pharmacists
   c. consumers
   d. anyone

Quick review question 39

An adverse event to which of the following should not be reported to MedWatch?

   a. St John's wort
   b. influenza vaccine
   c. valproic acid
   d. heparin

Quick review question 40

Which of the following is not a purpose of the VAERS program?

   a. asses the safety of newly licensed vaccines
   b. identify excipients used in various vaccines
   c. detect new, unusual, or rare vaccine adverse events
   d. identify vaccine lots with increased numbers or types of reported adverse events

Answers to quick review questions

1. B - An adverse event is any undesirable experience temporally associated with the use of a medical product. The word 'temporally' can be defined as relating to time.
2. D - A medication error is any incorrect or wrongful administration of a medication, such as a mistake in dosage or route of administration, failure to prescribe or administer the correct drug or formulation for a particular disease or condition, use of outdated drugs, failure to observe the correct time for administration of the drug, or lack of awareness of adverse effects of certain drug combinations. Option 'A' is the wrong patient, 'B' is the wrong route, and 'C' is the wrong dosage form (wrong dosage).
3. A - A risk evaluation and mitigation strategy (REMS) is a plan to manage a known or potential serious risk associated with a drug or biological product.
4. C - Since an error occurred that may have contributed to or resulted in permanent patient harm, it should be placed in the error, harm grouping.
5. A - Since this is a circumstance or event that has the capacity to cause error, but nothing incorrect actually happened, it should be placed in the no error grouping.
6. B - Since an error occurred, but the error did not reach the patient, it should be placed in the error, no harm grouping.
7. D - Since an error occurred that may have contributed to or resulted in the patient's death, it should be placed in the grouping error, death.
8. D - None of the examples follow best practices, as A uses 'µg' instead of 'mcg', B is missing a leading zero before the decimal '.025' instead of '0.025', and C uses the abbreviation 'qd' instead of 'daily'.
9. B - Zovirax and Doribax have near-homophony sounding names, which may cause someone to misinterpret which medication is needed.
10. A - High-alert medications are drugs that bear a heightened risk of causing significant patient harm when they are used in error.
11. C - A performance deficit is the failure to act in accordance

with education and training.

12. D - Antithrombotic agents, which would include warfarin, are considered high-alert medications.

13. C - Injectable antibiotics are not considered high-alert medications.

14. D - Most likely, the physician did not realize the increased likelihood of causing hearing loss when using gentamicin as a continuous infusion as opposed to giving it intermittently.

15. C - MS can mean morphine sulfate or magnesium sulfate and therefore can be misinterpreted - to avoid confusion the best practice is to write "morphine sulfate" or "magnesium sulfate".

16. A - High-alert medications are drugs that bear a heightened risk of causing significant patient harm when they are used in error.

17. A - The '5 rights' are right patient, right medication, right dose, right time, and right route. Sometimes, additional 'rights' are included such as right technique, right chart information, right documentation, and right attitude.

18. B - Tallman lettering (or Tall Man lettering) is the practice of writing part of a drug's name in upper case letters to help distinguish sound-alike, look-alike drugs from one another in order to avoid medication errors.

19. D - ISMP, FDA, The Joint Commission, and other safety-conscious organizations have promoted the use of tallman letters.

20. C - The barcode on the side of a manufacturer's bottle can be used to verify the NDC number.

21. D - Separating inventory can be a useful tool for preventing errors by making harder to grab the wrong product, whether it is a concern over incorrect dosage form, or an error caused by sound-alike, look-alike drugs being near each other on a shelf.

22. A - Tallman lettering (or Tall Man lettering) is the practice of writing part of a drug's name in upper case letters to help distinguish sound-alike, look-alike drugs from one another in order to avoid medication errors.

23. A - The Food and Drug Administration Amendments Act of 2007 authorizes the Food and Drug Administration (FDA) to require risk evaluation and mitigation strategies (REMS) on

medications, if necessary, to minimize the risks associated with some drugs.

24. D - As part of a drug utilization review, the following are areas that the pharmacist must screen: therapeutic duplication, drug–disease contraindications, drug–drug interactions, incorrect drug dosage, incorrect duration of treatment, drug–allergy interactions, and clinical abuse/misuse of medication.

25. D - As part of a drug utilization review, the following are areas that the pharmacist must screen: therapeutic duplication, drug–disease contraindications, drug–drug interactions, incorrect drug dosage, incorrect duration of treatment, drug–allergy interactions, and clinical abuse/misuse of medication.

26. A - As part of a drug utilization review, the following are areas that the pharmacist must screen: therapeutic duplication, drug–disease contraindications, drug–drug interactions, incorrect drug dosage, incorrect duration of treatment, drug–allergy interactions, and clinical abuse/misuse of medication.

27. A - Tallman lettering (or Tall Man lettering) is the practice of writing part of a drug's name in upper case letters to help distinguish sound-alike, look-alike drugs from one another in order to avoid medication errors.

28. B - The Food and Drug Administration (FDA) determines the requirements for patient package inserts.

29. D - A patient package insert or medication guide is a document provided along with a prescription medication to provide additional information about that drug. By providing this information to the patient, there is an expectation of improved therapeutic outcomes by improving compliance, and helping the patient to avoid some potential errors from medication misuse.

30. C - This is a drug-drug interaction as the aspirin will work in combination to increase the patient's anticoagulation to a potentially harmful level.

31. B - Technicians may assist in this process by identifying patients that could benefit from counseling and by gathering information, but all actual drug advice given to the patient should come from the pharmacist.

32. D - Medication therapy management (MTM), or an annual in-depth review of the medicare patient's medication profile is provided to add a safety feature to prevent adverse reactions and drug interactions, and to look at ways to reduce the patient cost.
33. B - MedWatch reporting is voluntary.
34. B - MedWatch is the Food and Drug Administration's reporting system for an adverse event.
35. C - The Vaccine Adverse Event Reporting System is a United States program for vaccine safety, co-managed by the Centers for Disease Control and Prevention (CDC) and the Food and Drug Administration (FDA).
36. C - You are required to report all significant adverse events that occur after vaccination of adults and children, even if you are not sure whether the vaccine caused the adverse event.
37. D - The National Medication Errors Reporting Program (MERP), MedWatch, and the Vaccine Adverse Event Reporting System (VAERS) are all error reporting systems.
38. D - Both patients and health care workers may submit an error report directly to MedWatch, although, patients are encouraged to take this form to their health care professional.
39. A - Medwatch may accept error reports for all medications (including OTC products and dietary supplements) excluding vaccines.
40. B - The primary objectives of VAERS are to: detect new, unusual, or rare vaccine adverse events (VAEs); monitor increases in known adverse events; identify potential patient risk factors for particular types of adverse events; identify vaccine lots with increased numbers or types of reported adverse events; and assess the safety of newly licensed vaccines.

CHAPTER 5 Pharmacy Quality Assurance

Key concepts

This chapter will cover the following knowledge areas to prepare you for the *Pharmacy Technician Certification Exam*:

- Quality assurance practices for medication and inventory control systems (e.g., matching National Drug Code (NDC) number, bar code, data entry)
- Infection control procedures and documentation (e.g., personal protective equipment [PPE], needle recapping)
- Risk management guidelines and regulations (e.g., error prevention strategies)
- Communication channels necessary to ensure appropriate follow-up and problem resolution (e.g., product recalls, shortages)
- Productivity, efficiency, and customer satisfaction measures

Terminology

To get started in this chapter, there are some terms that should be defined.

risk management - Risk management is intended to identify errors, assess the root cause(s), and implement procedures to reduce medication errors. Risk management also involves measuring outcomes to determine if successful, and then making additional process changes as necessary.

quality control (QC) - Quality control is the process of routinely checking/monitoring products to verify accuracy and/or appropriateness. Quality control typically looks at specific items at a given moment in time.

quality improvement - Quality improvement is the process of devising a method (quality assurance) to address concerns identified as potential sources of medication errors.

quality assurance (QA) - Quality assurance refers to the activities implemented so that requirements for a product or service will be fulfilled without error.

inventory - Inventory is simply the entire stock on hand for sale at a given time.

perpetual inventory - A perpetual inventory is a system that maintains a continuous count of every item in inventory so that it always shows the stock on hand. Some pharmacies maintain perpetual inventories on all products, while others may only do this with their schedule II medications.

reorder point - A reorder point provides minimum and maximum (a.k.a., par and max) stock levels which determine when a reorder is placed and for how much.

National Drug Code (NDC) - The National Drug Code is a unique product identifier used for medications intended for human use. It is a unique 10-digit, 3-segment numeric identifier assigned to each medication. The first segment identifies the manufacturer. The second segment identifies a specific strength, dosage form, and formulation for a particular manufacturer. The third segment identifies package forms and sizes.

lot number - A lot number is an identification number assigned to a particular quantity (or lot) of material from a single manufacturer made in a specific batch. Sometimes also called a batch number or control number.

The Joint Commission (TJC) - The Joint Commission, formerly the Joint Commission on Accreditation of Healthcare Organizations (JCAHO), is a nonprofit organization that accredits more than 20,000 healthcare organizations and programs in the United States.

hazardous drugs - Hazardous drugs are drugs that are known to cause genotoxicity, which is the ability to cause a change or mutation in genetic material; carcinogenicity, which is the ability to cause cancer in animal models, humans or both; teratogenicity, which is the ability to cause defects on fetal development or fetal malformation; and lastly, hazardous drugs are known to have the potential to cause fertility impairment, which is a major concern for most clinicians. These drugs can be classified as antineoplastics, cytotoxic agents, biologic agents, antiviral agents and immunosuppressive agents.

nosocomial infection - A nosocomial infection is an infection acquired while in the hospital.

personal protective equipment (PPE) - Personal protective equipment is worn by an individual to provide both protection to the wearer from the environment or specific items they are manipulating, and to prevent exposing the environment or the items being manipulated directly to the wearer of the PPE.

communication - Communication is the activity of providing information through the exchange of thoughts, messages, or information, as by speech, visuals, signals, writing, electronic means, or by behavior. This information is exchanged between two or more people.

Quick review question 1

Fill in the blank: _____ refers to the activities implemented so that requirements for a product or service will be fulfilled without error.

  a. Risk management
  b. Quality assurance
  c. Quality control
  d. Quality improvement

Quick review question 2

Which of the following terms is best described as a continuous count of merchandise on hand intended for resale?

  a. capital expenditure
  b. inventory
  c. perpetual inventory
  d. micromanagement

Quick review question 3

Which of the following is an identification number assigned to a particular quantity of material from a single manufacturer made in a specific batch?

a. lot number
b. NDC
c. expiration date
d. beyond use date

Quick review question 4

Which of the following situations is not an example of pharmacy communication?

a. a staff meeting where proper antibiotic dosing regimens are discussed
b. flagging a drug in the computer to let others know it is on back order
c. sending out a memo to inform the staff about a change in procedures
d. maintaining a personal list about third-party payor override codes

Risk management

The medical profession has a responsibility to provide safe, effective health care, which includes reducing medication errors. Risk management is the process of identifying errors, assessing the root cause(s), and implementing procedures to reduce medication errors. These three aspects of risk management can be termed respectively as quality control (QC), quality improvement, and quality assurance (QA). These errors will typically be identified through quality controls (QC) and spontaneous error reporting. Quality improvement involves reviewing trends to verify consistency and identify/assess areas that require changes. Quality assurance is the plan for any processes that need to be implemented to reduce errors. Risk management should also seek a means of quantification to determine if the quality improvement plan actually achieves its desired goals, often through QC.

Quality control

Quality control provides regular intervals in which specific items may be measured as correct or incorrect (i.e., the medication refrigerator was within proper range when checked this morning, or the vial of medication you filled was done correctly) to provide verification of the success of a particular quality assurance plan. This base line data is also necessary for future evaluations. Many healthcare facilities strongly encourage reporting of errors so they may gather information to improve or modify the process(es). To gather good information, facilities need to remove the stigma associated with an individual making an error by emphasizing shared responsibility for the organization, and promoting open discussion of errors at staff meetings.

Quality improvement

Quality improvement is the process of devising a method (quality assurance) to address concerns identified as potential sources of medication errors. There are many methods for quality improvement including: product improvement, process improvement, and people based improvement. Product improvement could be a redesign of the product or a labeling modification in such a way as to reduce the potential for misuse. Process improvement could involve removing potential sources of errors, or adding additional safety processes such as barcode scanning. People based improvement could be improving or modifying the training procedures of staff, or increasing the knowledge base of the consumer.

Quality assurance

Quality assurance (QA) refers to the activities implemented, often as a result of a quality improvement plan, so that requirements for a product or service will be fulfilled without error. This can be contrasted with QC, which is focused on specific process outputs. QC and error reporting is necessary to help assess how well a particular QA plan works.

Evaluating process changes

Often, quality improvement plans will need additional modifications after they are implemented, as the challenges of a complex work environment demonstrates the need for further changes. To achieve this, a mechanism will need to be included to obtain feedback. Feedback may come from QC statistics, error reporting, along with discussion involving staff and consumers. As the staff will have the majority of interactions with the processes, it is common for many institutions to adopt policies to regularly address medication and process errors at staff meetings.

Risk management is a continuous process that may require many incremental changes as you regularly measure, assess, evaluate, and manage your processes.

Quick review question 5

Which of the following is an example of risk management?

a. checking a medication to ensure it has been filled correctly before it leaves the pharmacy

b. having a policy that all high-alert medications must be checked by at least two pharmacists prior to dispensing

c. The pharmacy devises a method to make sure that parenteral nutrition solutions above 1000 mOsmol/L are never infused in a peripheral IV line.

d. Fluoxetine was accidentally repackaged incorrectly for the pharmacy robot as sulfamethoxazole and trimethoprim. The root cause was because the technician repackaged several medications at the same time, and the pharmacist did not realize multiple medications had been repackaged. The solution was to create a new policy that each drug that is repackaged must be set physically separate from each other prior to being checked.

Quick review question 6

Which of the following is an example of quality control?

a. checking a medication to ensure it has been filled correctly before it leaves the pharmacy

b. having a policy that all high-alert medications must be checked by at least two pharmacists prior to dispensing

c. The pharmacy devises a method to make sure that parenteral nutrition solutions above 1000 mOsmol/L are never infused in a peripheral IV line.

d. Fluoxetine was accidentally repackaged incorrectly for the pharmacy robot as sulfamethoxazole and trimethoprim. The root cause was because the technician repackaged several medications at the same time, and the pharmacist did not realize multiple medications had been repackaged. The solution was to create a new policy that each drug that is repackaged must be set physically separate from each other prior to being checked.

Quick review question 7

Which of the following is an example of quality improvement?

a. checking a medication to ensure it has been filled correctly before it leaves the pharmacy

b. having a policy that all high-alert medications must be checked by at least two pharmacists prior to dispensing

c. The pharmacy devises a method to make sure that parenteral nutrition solutions above 1000 mOsmol/L are never infused in a peripheral IV line.

d. Fluoxetine was accidentally repackaged incorrectly for the pharmacy robot as sulfamethoxazole and trimethoprim. The root cause was because the technician repackaged several medications at the same time, and the pharmacist did not realize multiple medications had been repackaged. The solution was to create a new policy that each drug that is repackaged must be set physically separate from each other prior to being checked.

Quick review question 8

Which of the following is an example of quality assurance?

a. checking a medication to ensure it has been filled correctly before it leaves the pharmacy

b. having a policy that all high-alert medications must be checked by at least two pharmacists prior to dispensing

c. the pharmacy devises a method to make sure that parenteral nutrition solutions above 1000 mOsmol/L are never infused in a peripheral IV line

d. Fluoxetine was accidentally repackaged incorrectly for the pharmacy robot as sulfamethoxazole and trimethoprim. The root cause was because the technician repackaged several medications at the same time and the pharmacist did not realize multiple medications had been repackaged. The solution was to create a new policy that each drug that is repackaged must be set physically separate from each other prior to being checked.

Inventory control

Inventory control can be used to provide quality assurance in a number of ways including: ensuring that the necessary medications are on hand to provide therapy, establishing methods to ensure the proper medications are being dispensed, removing expired and short-dated medications from stock, regularly checking for drug recalls and performing all appropriate follow-ups, and segregating/separating inventory to minimize the potential for miss-picks. Properly trained staff, careful attention to detail, and automation are important to proper and safe inventory control.

Inventory levels

Most pharmacies establish maximum and minimum inventory levels to try and assure an adequate, but not excessive, quantity of medicines are on the shelf. Often, max and min levels are monitored by computer systems that maintain perpetual inventories, but pharmacy staff will need to initially establish those levels, and may occasionally need to intervene and adjust those levels. Furthermore, some pharmacies still do not have automated inventory systems. When you establish a max level, that is the most you want to have on the shelves, and you should not exceed that quantity. When you establish a minimum level (or par level), that is the point when you order more (commonly called your reorder point), but not till you reach that minimum quantity. By controlling the inventory level, it will help with both order fulfillment and minimize the potential for medications to sit around until they expire.

If your perpetual inventory is capable of tracking NDCs, lot numbers, and expiration dates, it can help you monitor not just for inventory levels, but also for expired medications, drug recalls, and can even make sure that your pharmacy management software is using available drugs (based on NDC) for orders that are being entered by the staff.

Order entry and barcode scanning

As technology in the pharmacy becomes increasingly prevalent, it provides additional safety checks within the dispensing process. The order entry process can be automatically linked to the NDCs stocked by a particular pharmacy; then, when the medication label is printed, it may have a barcode linking it to a patient profile and a particular medication. A barcode scanner can be used to verify that the correct drug is being used to fill the patient prescription. In many institutional settings, patients are also provided with barcodes on their hospital bracelets. This further allows those providing direct patient care to ensure that it is the right drug, going to the right patient, at the right time.

Removing medication from stock

Pharmacies will often need to remove medications from stock, whether it is due to the medications being expired, short-dated (near their expiration), or recalled.

Proper rotation of inventory and periodic checking of expirations (pulling both expired and short-dated medications) help to reduce the potential for dispensing expired drugs. It also maximizes the utilization of inventory before medications become outdated. When looking at expirations on medication vials, it is important to note that if a medication only mentions the month and year but not the day, then you are to treat it as expiring at the end of the month. As an example, if a medication is marked as expiring on 02/2020, then you would treat it as expiring on February 29, 2020.

Sometimes, medications may need to be recalled for a variety of reasons. There are three drug recall classifications: class I, class II, and class III.

- Class I drug recalls have reasonable probability that the use of, or exposure to, a violative product will cause serious adverse health consequences or death. An example would include the diet aid Fen-Phen (fenfluramine/phentermine), which caused irreparable heart valve damage and pulmonary hypertension.

- Class II drug recalls involve the use of, or exposure to, a violative product that may cause temporary or medically reversible adverse health consequences, or where the probability of serious adverse health consequences is remote. An example would include a medication that is under-strength, but that is not used to treat life-threatening situations.
- Class III drug recalls involve medications that the use of, or exposure to, a violative product is not likely to cause adverse health consequences. Examples might be a container defect (plastic material delaminating or a lid that does not seal), off-taste or incorrect color, and simple text errors such as the incorrect expiration date on the manufacturer's bottle.

Pharmacies will regularly be notified about drug recalls through both the FDA and the pharmacy's wholesale supplier. Often, recalls affect just specific batches or runs of medications. If this is the case, the recall will include the specific lot number(s) being recalled. Depending on the seriousness of the recall, the pharmacy may even be required to contact patients that potentially received these recalled drugs.

Medications removed from stock should be carefully stored to avoid accidentally dispensing these medications. Some medications may be able to be returned to the wholesaler or manufacturer for credit, controlled substances may need to be disposed of through DEA approved means (contact your local DEA office for specific advice), and other medications may be left to the pharmacy to find an appropriate means for drug disposal.

Separating inventory

Segregating inventory by drug categories helps to prevent potentially harmful errors. Examples include internal versus external products, hazardous drugs, and volatile or flammable substances.

The Joint Commission (TJC), formerly known as the Joint Commission on Accreditation of Healthcare Organizations (JCAHO), requires that internal and external medications must be stored separately. This reduces the potential that someone will dispense or administer an external product for internal use. The Joint Commission also has requirements for separate storage of oncology drugs and volatile or flammable substances.

Hazardous drugs (i.e., oncology drugs) should have a separate space on the shelves and be labeled in such a way that it will alert staff of the hazardous potential of these medications. Oncology drugs are often cytotoxic themselves, and must be handled with extreme care. They should be received in a sealed protective outer bag that restricts dissemination of the drug if the container leaks or is broken. When potential exists for exposure to hazardous drugs, all personnel involved must wear appropriate personal protective equipment while following a hazardous materials cleanup procedure. All exposed materials must be properly disposed of in hazardous waste containers.

Volatile or flammable substances (including tax free alcohol) require careful storage. They must have a cool location that is properly ventilated. Their storage area must be designed to reduce fire and explosion potential.

Quick review question 9

Tracking medication lot numbers may be useful for which of the following?

a. tracking expiration dates
b. using barcode scanning to ensure the proper medication is being dispensed
c. checking for drug recalls
d. none of the above

Quick review question 10

Why might a pharmacy want to remove a medication from stock?

a. expired
b. short dated
c. drug was recalled
d. all of the above

Quick review question 11

Which medications do The Joint Commission specifically require to be stored separately?

a. internal and external products
b. sugar free products and products with sugar
c. birth control and fertility drugs
d. all of the above

Quick review question 12

How should volatile or flammable substances be stored?

a. dispersed throughout the shelves with various legend drugs
b. in a storage area designed to reduce fire and explosion potential
c. They should be stored with other hazardous drugs (i.e., antineoplastic agents and antivirals).
d. Pharmacies should never carry volatile or flammable substances.

Infection control

Infection control is concerned with preventing nosocomial or healthcare-associated infection. Infection control addresses factors related to the spread of infections within the healthcare setting (whether patient-to-patient, from patients to staff and from staff to patients, or among staff), including prevention via hand hygiene, and through the use of personal protective equipment (PPE). Also, with the rise in pharmacies providing injections, there is also an increased concern of infection from contaminated needles.

When assembling compounded sterile preparations, preparers should don personal protective equipment (PPE) to further minimize the likelihood that they will contaminate the final product. USP 797 provides requirements for how personnel should wash and garb, including the order that they should don their PPE, from dirtiest to cleanest. First, personnel should remove unnecessary outer garments and visible jewelry (scarves, rings, ear rings, etc.). Next, they should don shoe covers, facial hair and hair covers, and masks (optionally, they may include a face shield). Next, they must perform appropriate hand hygiene.

USP 797 states:

"...personnel perform a thorough hand-cleansing procedure by removing debris from under fingernails using a nail cleaner under running warm water followed by vigorous hand and arm washing to the elbows for at least 30 seconds with either nonantimicrobial or antimicrobial soap and water."

After hand washing, a nonshedding gown with sleeves that fit snugly around the wrists should be put on. Once inside the buffer area, an antiseptic hand cleansing using a waterless alcohol-based hand scrub should be performed. Sterile gloves are the last item donned prior to compounding.

Recapping needles at appropriate times can decrease opportunities for unintended needle sticks. Any time a syringe is being drawn up, it should be recapped if it is not going to be used immediately. Also, the old practice of breaking used needles prior to discarding them into a sharps container is strongly discouraged, as the process of breaking the needle is another opportunity in which someone could incur an accidental needle stick.

Quick review question 13

Infection control policies should address spread of disease between which group(s)?

a. patient-to-patient
b. from patients to staff and from staff to patients
c. among staff
d. all of the above

Quick review question 14

What should be done with a syringe needle if it is not going to be used immediately?

a. the needle should be discarded
b. the needle should be broken off
c. the needle should be recapped
d. the entire syringe should be placed in a sharps container

Quick review question 15

When preparing to don PPE, what should be done with visible jewelry?

a. it should be covered by your PPE
b. it should be removed
c. it should be coveted
d. there are no specific guidelines with respect to jewelry

Quick review question 16

When donning PPE, when should you use a waterless alcohol based hand scrub?

a. prior to donning gloves
b. prior to washing hands
c. in between each CSP
d. you should never use an alcohol based hand scrub

Communications

Communication channels are necessary within the pharmacy, interprofessionally, and between the pharmacy and its patients to help assure good therapeutic outcomes.

Within the pharmacy itself, pharmacy staff should be sharing information about drug recalls, drug shortages and back orders, along with important patient specific information. Quality assurance plans should address all these situations to reduce potential for frustration and errors. This information may be disseminated throughout the staff in several ways, including emails, printed memos, bulletin boards and white boards for posting information, and even setting up information within the pharmacy management software to help flag critical items (be careful about sharing sensitive information by means that may allow to much public access to the information). It is also advisable to discuss these situations at staff meetings to ensure everyone is aware of the potential challenges and what circumventions may be necessary to handle these situations.

Interprofessionally, pharmacists and pharmacy technicians will often need to contact physicians, nurses, third-party payors, and other professional groups and individuals in order to provide quality care. These communications may include attempting to get approval for a medication, needing a prescription clarification, concerns over potential interactions, managing third-party payor claim rejections, or just making sure that your patient profile is complete and accurate. Some of the processes may be automated, or scripted, but other items that require more clinical judgment may need to be reserved for pharmacists.

Communications between the pharmacy staff and the patients is a critical point as pharmacies have the ability to both gather and disseminate information for the patient. During the in-take process, the pharmacy should gather information from all the various care-providers that the patient may see, ailments the patient is being treated for, and allergies. The pharmacy also has the ability to include information about dietary supplements and over the counter medications the patient may use. The collection of this data should be a routine process within the pharmacy in order to screen for interactions and contraindications. The pharmacy should disseminate information by offering counseling, providing medication guides, and assuring that all appropriate auxiliary labels are attached to prescriptions. Also, patients should be informed when there are problems with or changes to their medication therapies so they can be compliant with their therapy.

As an example of communication within the pharmacy, interprofessionally, and between the pharmacy and patients, in early 2009 the FDA posted a recall on metoprolol succinate from a specific manufacturer. The FDA alert would have gone out to anyone that had signed up for it at www.fda.gov/Safety/Recalls, and the next pharmacy delivery would have included the appropriate recall information. Upon receipt of the recall information, the pharmacy staff should have pulled any offending bottles from the shelves, and if deemed appropriate, contacted patients at home about the significant medication problem. The pharmacy should have then posted the recall information for the staff to be aware of this important information. Also, due to the widespread nature of this particular recall, that pharmacy needed to make all the staff aware of the drug shortage caused by this. In the short term, any patients that needed metoprolol succinate needed to either be switched to a different medication by their physician (which often occurred after the pharmacy contacted the physician about the shortage), and find a substitution that was covered by the patient's insurance coverage when applicable. Also, the pharmacy had a responsibility to inform the patient about their medication change.

Quick review question 17

Pharmacies often have communications with which group(s)?

a. physicians
b. patients
c. third-party payors
d. all of the above

Quick review question 18

Why would a pharmacy need to contact a physician?

a. for clarification on a prescription
b. due to concerns over an interaction
c. to inform a physician about an insurance claim rejection
d. all of the above

Quick review question 19

Communication within the pharmacy should include all of the following except which of the following?

a. drug recalls
b. gossip
c. drug shortages
d. back orders

Quick review question 20

During the in-take process, which of the following is not considered critical?

a. the concurrent use of any OTC and/or dietary supplements
b. sexual preference
c. allergies
d. various conditions the patient is being treated for

Quick review question 21

Which of the following is an example of information that should be routinely disseminated to patients?

a. offering counseling
b. medication guides

c. appropriate auxiliary labels on medications
d. all of the above

Productivity

Through the use of automation, and by establishing routines for how procedures should be done, quicker turn around times can be offered for patients with fewer errors, which has the potential to improve customer satisfaction. Automation to aid these processes may include items like barcode scanners, electronic pill counters, and automated filling cabinets (i.e., *ScriptPro*). These products allow for some steps to be made easier and free up the technicians to work on other tasks. Also, by establishing routines, it is easier to keep track of where you are at in a particular process when the inevitable work flow interruptions occur. This allows the staff to more quickly determine where it is at in its process when they return to, it and minimizes the potential of either skipping or repeating a step. Some pharmacy management software systems on the market today even offer assistance with this by requiring a barcode scan as each part of the filling process is completed. The desire of this increased productivity is not just intended to provide improved abilities to handle larger workloads with fewer errors but, also to help free up pharmacy staff to spend more time assisting the patients with their questions, concerns, and needs.

Quick review question 22

Which of the following is not a potential benefit from the improved efficiencies that often result from pharmacies adopting automation and establishing procedural routines?

a. staff may have more time to gossip
b. pharmacies may be capable of offering additional services as a result of less time being required by other processes
c. more time is available to provide counseling for the patients, if they desire it
d. the patients often don't need to wait as long for prescriptions to be filled

Quick review question 23

Which of the following is not an example of a pharmacy automation that can improve productivity?

a. an electronic tablet counter
b. barcode scanners
c. automated filling cabinets
d. double data entry

Quick review question 24

Which of the following may improve productivity and quality assurance?

a. establishing routines
b. pharmacy automation
c. both a and b
d. none of the above

Answers to quick review questions

1. B - Quality assurance refers to the activities implemented so that requirements for a product or service will be fulfilled without error.
2. C - A perpetual inventory is a system that maintains a continuous count of every item in inventory so that it always shows the stock on hand.
3. A - A lot number is an identification number assigned to a particular quantity (or lot) of material from a single manufacturer made in a specific batch. Sometimes also called a batch number or control number.
4. D - Because communication requires information to be exchanged between two or more people personal notes and lists that are not shred do not count as communications.
5. D - Risk management is intended to identify errors, assess the root cause(s), and implement procedures to reduce medication errors.
6. A - Quality control is the process of routinely checking/monitoring products to verify accuracy and/or appropriateness. Quality control typically looks at specific items at a given moment in time.
7. C - Quality improvement is the process of devising a method

(quality assurance) to address concerns identified as potential sources of medication errors.

8. B - Quality assurance refers to the activities implemented so that requirements for a product or service will be fulfilled without error.

9. C - Often recalls affect just specific batches or runs of medications. If this is the case, the recall will include the specific lot number(s) being recalled.

10. D - Pharmacies will often need to remove medications from stock, whether it is due to the medications being expired, short-dated (near their expiration), or recalled.

11. A - The Joint Commission (TJC) requires that internal and external medications must be stored separately. This reduces the potential that someone will dispense or administer an external product for internal use.

12. B - Volatile or flammable substances (including tax free alcohol) require careful storage. They must have a cool location that is properly ventilated. Their storage area must be designed to reduce fire and explosion potential.

13. D - Infection control addresses factors related to the spread of infections within the healthcare setting (whether patient-to-patient, from patients to staff and from staff to patients, or among staff), including prevention via hand hygiene, and through the use of personal protective equipment (PPE).

14. C - Any time a syringe is being drawn up, it should be recapped if it is not going to be used immediately.

15. B - When preparing to don PPE personnel should remove unnecessary outer garments and visible jewelry (scarves, rings, ear rings, etc.).

16. A - When donning PPE, personnel should first remove unnecessary outer garments and visible jewelry (scarves, rings, ear rings, etc.). Next, they should don shoe covers, facial hair and hair covers, and masks (optionally they may include a face shield). Next, they must perform appropriate hand hygiene. After hand washing, a nonshedding gown with sleeves that fit snugly around the wrists should be put on. Once inside the buffer area, an antiseptic hand cleansing using a waterless alcohol-based hand scrub should be performed. Sterile gloves are the last item donned prior to

compounding.

17. D - Communication channels are necessary (and routine) within the pharmacy (pharmacists and technicians), interprofessionally (physicians, nurses, third-party payors, etc.), and between the pharmacy and its patients to help assure good therapeutic outcomes.

18. D - Interprofessionally, pharmacists and pharmacy technicians will often need to contact physicians, nurses, third-party payors, and other professional groups and individuals in order to provide quality care. These communications may include attempting to get approval for a medication, needing a prescription clarification, concerns over potential interactions, dealing with third-party payor claim rejections, or just making sure that your patient profile is complete and accurate.

19. B - Within the pharmacy itself, pharmacy staff should be sharing information about drug recalls, drug shortages and back orders, along with important patient specific information.

20. B - During the in-take process, the pharmacy should gather information from all the various care-providers that the patient may see, ailments the patient is being treated for, and allergies, The pharmacy also has the ability to include information about dietary supplements and over the counter medications the patient may use.

21. D - The pharmacy should disseminate information to patients by offering counseling, providing medication guides, and assuring that all appropriate auxiliary labels are attached to prescriptions. Also, patients should be informed when there are problems with or changes to their medication therapies so they can be compliant with their therapy.

22. A - While some down time may be desired for staff, observing the pharmacy staff gossiping does not provide a tangible benefit for the patient.

23. D - Double data entry is the idea of entering information into a computer system multiple times to ensure that the correct information was entered previously. While some industries use this as a quality control, entering information multiple times does not improve productivity.

24. C - The establishment of routines for various procedures,

and the use of automation both have the potential to improve productivity and quality assurance.

CHAPTER 6 Medication Order Entry and Fill Process

Key concepts

This chapter will cover the following knowledge areas to prepare you for the *Pharmacy Technician Certification Exam*:

- Order entry process
- Intake, interpretation, and data entry
- Calculate doses required
- Fill process (e.g., select appropriate product, apply special handling requirements, measure, and prepare product for final check)
- Labeling requirements (e.g., auxiliary and warning labels, expiration date, patient specific information)
- Packaging requirements (e.g., type of bags, syringes, glass, PVC, child resistant, light resistant)
- Dispensing process (e.g., validation, documentation and distribution)

Terminology

To get started in this chapter, there are some terms that should be defined.

prescription origin code (POC) - Prescription origin codes (POC) were first created by the Centers for Medicare and Medicaid Services (CMS) to help gather data on where prescriptions come from (0 = Unknown, 1 = Written, 2 = Telephone, 3 E-prescription, 4 = Fax). POCs are now mandatory for both Medicare and Medicaid patients and many other insurance companies require this as well.

e-prescribing - E-prescribing is the computer-based electronic generation, transmission, and filling of a medical prescription, taking the place of paper and faxed prescriptions. E-prescribing allows a physician, nurse practitioner, or physician assistant to electronically transmit a new prescription or renewal authorization to a community or mail-order pharmacy.

pharmacy benefits manager (PBM) - A PBM is a company that acts as an intermediary between the pharmacy and the insurance plan.

computerized prescriber order entry (CPOE) - Computerized prescriber order entry, also called computerized physician order entry, is a process of electronic entry of practitioner instructions for the treatment of patients under their care. CPOE systems are used for processing orders in institutional settings.

superscription - The superscription consists of the heading on a prescription where the symbol Rx is found. The Rx symbol comes before the inscription.

inscription - The inscription is also called the body of the prescription, and provides the names and quantities of the chief ingredients of the prescription. Also in the inscription, you find the dose and dosage form, such as tablet, suspension, capsule, or syrup.

signatura - The signatura (also called sig, or transcription), gives instructions on a prescription to the patient on how, how much, when, and how long the drug is to be taken. These instructions are preceded by the symbol "S" or "Sig." from the Latin, meaning "write" or "label." Whenever translating the signatura into instructions for a patient, begin it with an action verb such as take, inhale, spray, inject, place, swish, or whatever other verb seems appropriate for the medication.

bank identification number (BIN) - On a health insurance card, a BIN is a six digit number used to identify a specific plan from a carrier making it easier for the PBM to process your prescription online. No actual bank is involved in this part of the process; the name is a hold over from early electronic banking terminology.

dispense as written (DAW) codes - DAW codes are used to

provide a quick explanation of whether or not a generic version of the medication is allowed to be dispensed, and if not, then why and whom deemed the brand name product to be necessary. A DAW code of '0' applies to most prescriptions as they allow for generic substitution, and patients are generally willing to receive the more affordable version. If a physician requires a specific medication to be dispensed, they will typically note this on the prescription. This is considered a DAW code of '1'. Sometimes a patient may request that they receive a brand name product even if a prescriber allowed for generic substitution. This would be classified as a DAW code of '2'. Other DAW codes are less frequently used. The following is a succinct list of the other DAW codes: 3 = substitution allowed - pharmacist selected product dispensed; 4 = substitution allowed - generic drug not in stock; 5 = substitution allowed - brand drug dispensed as generic; 6 = override; 7 = substitution not allowed - brand drug mandated by law; 8 = substitution allowed - generic drug not available in marketplace; and 9 = other.

Quick review question 1

E-prescribing may be used by which group to send prescription to the pharmacy?

a. physicians
b. nurse practitioners
c. physician assistants
d. all of the above

Quick review question 2

CPOE is primarily used in which practice settings?

a. community pharmacy
b. mail order pharmacy
c. hospitals
d. all of the above

Quick review question 3

Which part of a prescription provides the instructions for the patient?

a. inscription
b. subscription
c. signatura

d. none of the above

Quick review question 4

What is a PBM?

a. an intermediary between the pharmacy and the insurance plan
b. a meaningless acronym based on old banking term
c. a type of bowel movement
d. none of the above

Quick review question 5

What does the acronym 'BIN', as commonly found on a health insurance card, stand for?

a. benefits identification number
b. bank identification number
c. business identification number
d. none of the above

Quick review question 6

Typically, when both the patient and the physician allow for generic substitution, which DAW code should be used?

a. DAW 0
b. DAW 1
c. DAW 2
d. DAW 3

Quick review question 7

When a physician determines that a brand name product is medically necessary, which DAW code should be used?

a. DAW 0
b. DAW 1
c. DAW 2
d. DAW 3

Prescription intake

Prescription intake, which is often the responsibility of a pharmacy technician, involves both the receipt of the initial prescription and, in a community pharmacy setting, gathering pertinent patient data.

Receiving the prescription

With respect to prescription intake, the first thing to address is the means by which the medication itself arrives in the pharmacy. Prescriptions may arrive on a traditional prescription form (there are specific security requirements on this if the patient is using Medicaid), a fax or phone call from an appropriately licensed prescriber, e-prescribing in an outpatient setting, and computerized prescriber order entry (CPOE) in an institutional setting. Each of these means have various federal regulations associated with them, and individual states may place additional restrictions on how they are used.

When a community pharmacy receives a prescription it is a requirement to note the source of the prescription if it is for a Medicare or Medicaid patient. As many individual insurance companies also require this information it has become a common practice for community pharmacies to track where all prescriptions come from. These prescriptions are tracked using prescription origin codes (POC) which are commonly entered into the pharmacy management software. The prescription origin codes are as follows:

0 = Unknown, this is used when the manner in which the original prescription was received is not known, which may be the case in a transferred prescription
1 = Written prescription via paper which includes computer printed prescriptions that a physician signs as well as tradition prescription forms
2 = Telephone prescription obtained via oral instruction or interactive voice response
3 = E-prescriptions securely transferred from a computer to the pharmacy
4 = Facsimile prescription obtained via fax transmission including an e-Fax where a scanned image is sent to the pharmacy and either printed or displayed on a monitor/screen

A written prescription should contain the following information at a minimum:

- the prescriber's name, address, and telephone number,
- if the order is for a controlled substance, the prescriber's DEA number,
- the patient's name, the date of issuance,
- the name of the medication or device prescribed and dispensing instructions (if necessary),
- the directions for the use of the prescription,
- any refills (if authorized),
- special labeling and other instructions, and
- the prescriber's signature.

The Medicaid Tamper-Resistant Prescription Pad Law has placed additional requirements on written prescriptions to help ensure the legitimacy of the prescriptions being received in the pharmacy. Since October 1, 2008, all Medicaid scripts must contain one or more industry recognized features from each of three categories of security as specified by the Centers for Medicare and Medicaid Service (CMS).

Category One - One or More industry-recognized features designed to prevent unauthorized copying of a completed or blank prescription form.

- Void Pantograph background (Hidden Message Technology)
- Reverse Rx Symbol
- Micro Printing
- Artificial Watermark on back of script
- Coin Activated Ink

Category Two - One or more industry-recognized features designed to prevent the erasure or modification of information written on a prescription by the prescriber.

- Colored Shaded Pantograph background
- Toner Grip Security Coating
- "Check and Balance" printed features such as "quantity" check boxes, and space to indicate "number of medications" written on prescription form

Category Three - One or more industry-recognized features designed to prevent the use of counterfeit prescription forms.

- Security Feature Warning Box and Warning Bands
- Security Back Printing
- Coin Activated Validation
- Batch Number identification
- Secure Rub Color Change Ink
- Consecutive Numbering

Physicians may, and often do, use these tamper-resistant prescriptions for their other patients as well.

Prescribers may send prescriptions to the pharmacy via a fax machine. The same requirements listed on written prescriptions apply for faxed prescriptions (although prescribers do not need to use tamper-resistant prescription pads for faxed orders), but there is an additional limitation. Schedule II medications may not be faxed under ordinary circumstances.

DEA has granted three exceptions to the fax prescription requirements for Schedule II controlled substances. The fax of a Schedule II prescription may serve as the original prescription as follows:

- A practitioner prescribing Schedule II narcotic controlled substances to be compounded for the direct administration to a patient by parenteral, intravenous, intramuscular, subcutaneous or intraspinal infusion may transmit the prescription by fax. The pharmacy will consider the faxed prescription a "written prescription" and no further prescription verification is required. All normal requirements of a legal prescription must be followed.
- Practitioners prescribing Schedule II controlled substances for residents of Long Term Care Facilities (LTCF) may transmit a prescription by fax to the dispensing pharmacy. The practitioner's agent may also transmit the prescription to the pharmacy. The fax prescription serves as the original written prescription for the pharmacy.
- A practitioner prescribing a Schedule II narcotic controlled substance for a patient enrolled in a hospice care program certified and/or paid for by Medicare under Title XVIII, or a hospice program which is licensed by the state may transmit a prescription to the dispensing pharmacy by fax. The

practitioner (or their agent) may transmit the prescription to the pharmacy. The practitioner will note on the prescription that it is for a hospice patient. The fax serves as the original written prescription.

A phone order prescription (also called a verbal order), should include all the same information as a written prescription as the pharmacist will need to reduce it to writing to later be filed with the other prescriptions. Under ordinary circumstances, Schedule II medications may not be called in.

For Schedule II controlled substances, an oral order is only permitted in an emergency situation. An emergency situation is defined as a situation in which:

- Immediate administration of the controlled substance is necessary for the proper treatment of the patient.
- No appropriate alternative treatment is available.
- Provision of a written prescription to the pharmacist prior to dispensing is not reasonably possible for the prescribing physician.

In an emergency, a practitioner may call in a prescription for a Schedule II controlled substance by telephone to the pharmacy, and the pharmacist may dispense the prescription provided that the quantity prescribed and dispensed is limited to the amount adequate to treat the patient during the emergency period. The prescribing practitioner must provide a written and signed prescription to the pharmacist within seven days. Further, the pharmacist must notify DEA if the prescription is not received.

E-prescribing has become a very common practice. Since 2007, pharmacies have been allowed to transmit prescriptions electronically using properly certified software (i.e., SureScripts) for most legend drugs. As of June 1, 2010, physicians and pharmacies are also allowed to transmit prescriptions for Schedule II, III, IV, and V medications. While this is a recent shift in federal law, some states may still prohibit e-prescribing for controlled substances.

Through the requirement of using computer programs that communicate through SureScripts to electronically prescribe outpatient prescriptions, SureScripts has the responsibility of authenticating both the receipt and delivery of the electronic prescription. E-prescribing also allows the prescriber to verify whether or not a particular medication is covered by the patient's pharmacy benefits manager (PBM).

Institutional settings are more likely to utilize CPOE (computerized prescriber order entry, also called computerized physician order entry) than e-prescribing. CPOE is a process of electronic entry of practitioner instructions for the treatment of patients under their care. These orders are communicated over a computer network to the medical staff, or to the departments (pharmacy, laboratory, or radiology) responsible for fulfilling the order. CPOE decreases delay in order completion, reduces errors related to handwriting or transcription, allows order entry at the point of care or off-site, provides error-checking for duplicate or incorrect doses or tests, and simplifies inventory and posting of charges.

Gathering patient data

When a patient first arrives at a community pharmacy, the pharmacy team will need to gather/verify various important pieces of patient data including:

- gather drug and disease information,
- ensure that the pharmacy has the correct name, address, contact information, and any other pertinent data,
- document/update allergy information, and
- verify/update medication insurance information.

In an inpatient setting, while a pharmacy will still want all this information to be completed, it is typically the responsibility of other departments.

Quick review question 8

Which of the following would be an appropriate means for receiving a prescription in a community pharmacy?

a. a prescription form
b. a fax or phone call from a physician
c. an e-prescription
d. all of the above

Quick review question 9

If a prescriber uses a prescription form to order a controlled substance, what must appear on the form?

   a. their DEA number
   b. their NPI
   c. their medical license
   d. all of the above

Quick review question 10

For which group of patients must prescribers use tamper-resistant prescriptions?

   a. all patients
   b. Medicaid patients
   c. Medicare patients
   d. Tamper-resistant prescriptions are not required for anyone.

Quick review question 11

For which group of patients may prescribers use e-prescriptions?

   a. all patients
   b. Medicaid patients
   c. Medicare patients
   d. E-prescriptions are still in the proposal phase and can not be used by anyone yet.

Quick review question 12

Which company has the responsibility of authenticating e-prescriptions?

   a. Each community pharmacy has the responsibility to authenticate these prescriptions themselves.
   b. McKesson
   c. SureScript
   d. No company needs to authenticate e-prescriptions.

Quick review question 13

Which of the following is not an advantage of CPOE?

   a. reduces errors related to hand writing

b. the low up-front costs to implementing CPOE

c. provides error-checking for duplicate or incorrect doses or tests

d. simplifies inventory and posting of charges

Quick review question 14

During prescription intake what patient data should the pharmacy gather/verify?

a. patient drug, disease, and allergy information

b. the patient's address and contact information

c. medication insurance information

d. all of the above

Prescription translation

Once the prescription has entered the pharmacy, it becomes the responsibility of the pharmacy staff to decipher and fill the medication for the patient. This will likely require the translation of various medical abbreviations and some calculation to ensure that the proper quantity is being dispensed.

Abbreviations

Prescriptions have been obfuscated by a combination of Latin and English abbreviations (sometimes they even throw in Greek words). They are commonly used on prescriptions to communicate essential information on formulations, preparation, dosage regimens, and administration of the medication. There are approximately 20,000 medical abbreviations; instead of providing an exhaustive and meaningless list, this section will focus on the most common medical abbreviations that are necessary for interpreting prescriptions and performing calculations.

The following lists are broken into five categories including route, dosage form, time, measurement, and a catch all category simply named "other." The abbreviations can often be written with or without the 'periods' and in upper or lower case letters (e.g., p.o. and PO both mean 'by mouth'). The format on these lists will be to provide the abbreviation, followed by its intended meaning.

Route

aa - affected area
a.d. - right ear
a.s. - left ear
a.u. - each ear
IM - intramuscular
IV - intravenous
IVP - intravenous push
IVPB - intravenous piggyback
KVO - keep vein open
n.g.t. - naso-gastric tube
n.p.o. - nothing by mouth
nare - nostril
o.d. - right eye
o.s. - left eye
o.u. - each eye
per neb - by nebulizer
p.o. - by mouth
p.r. - rectally
p.v. - vaginally
SC or SQ - subcutaneously
S.L. - sublingually (under the tongue)
top. - topically

Some additional notes on these routes of administration are necessary. The abbreviation 'a.d.' if written without periods, 'ad', can also mean 'to' or 'up to'. Also, subcutaneously can be abbreviated as 'SC' or 'SQ'. While amongst health care professionals we would use the phrase sublingual as a route of administration, it may be necessary to translate 'SL' as 'under the tongue' for many patients.

Dosage form

amp. - ampule
aq or aqua - water
caps - capsule
cm or crm - cream
elix. - elixir
liq. - liquid
sol. - solution
supp. - suppository

SR, XR, *or* XL - slow/extended release
syr. - syrup
tab. - tablet
ung. *or* oint - ointment

The abbreviation 'cm' can be translated as either 'cream' or 'centimeter'. Use context clues from the rest of the prescription to determine which translation is appropriate.

Time or how often

a.c. - before food, before meals
a.m. - morning
atc - around the clock
b.i.d. *or* bid - twice a day
b.i.w. *or* biw - twice a week
h *or* ° - hour
h.s. - at bedtime
p.c. - after food, after meals
p.m. - evening
p.r.n. *or* prn - as needed
q.i.d. *or* qid - four times a day
q - each, every
q.d. - every day
q_h *or* q_° - every__hour(s) (i.e., q8h would be translated as every 8 hours)
qod - every other day
stat - immediately
t.i.d. *or* tid - three times a day
t.i.w. *or* tiw - three times a week
wa - while awake

Measurement

i, ii, ... - one, two, etc. (often Roman numerals will be written on prescriptions using lowercase letters with lines over top of them)
ad - to, up to
aq. ad - add water up to
BSA - body surface area
cc - cubic centimeter
dil - dilute
f *or* fl. - fluid

fl. oz. - fluid ounce
g, G, *or* gm - gram
gr. - grain
gtt - drop(s)
l *or* L - liter
mcg *or* µg - microgram
mEq - milliequivalent
mg - milligram
ml *or* mL - milliliter
q.s. - a sufficient quantity
q.s. ad - add a sufficient quantity to make
ss - one-half (commonly used with Roman numerals to add a value of 0.5)
Tbs *or* T - tablespoon
tsp *or* t - teaspoon
U - unit
> - greater than
< - less than

Other

c - with
disp. - dispense
n/v - nausea and vomiting
neb - nebulizer
NR - no refills
NS *or* NSS - normal saline, normal saline solution
s - without
Sig *or* S - write, label
SOB - shortness of breath
T.O. - telephone order
ut dict *or* u.d. - as directed
V.O. - verbal order

Parts of a prescription

Traditionally, a prescription is a written order for compounding, dispensing, and administering drugs to a specific client or patient, and once it is signed by the physician, it becomes a legal document. Prescriptions are required for all medications that require the supervision of a physician (those that must be controlled because they are addictive and carry the potential of being abused, and those that could cause health threats from side effects if taken incorrectly - for example: cardiac medications, controlled substances, and antibiotics).

The following is a list of the parts of a prescription:

- *Patient Information*, which may include information such as name, address, age, weight, height, and allergies.
- *Superscription*, which is the 'Rx' symbol that we typically translate as, "Take thus."
- *Inscription*, which is the actual medication or compounding request.
- *Subscription*, or how much to dispense.
- *Signatura*, which is the instruction set intended for the patient.
- *Date*, which is when the prescription was written. Prescriptions for medications and supply that are not considered controlled substances are good for up to 1 year from when the prescription was written.
- *Signature lines*, which is where the prescriber provides their signature and indicates their degree. Often, this is also where a prescriber may indicate their preferences with regard to generic substitution.
- *Prescriber information*, which includes the physician's name, practice location address, telephone number and fax number. This may also include the prescriber's NPI number and appropriate license numbers.
- *DEA#*, DEA numbers are required for controlled substances.
- *Refills*, which simply indicates how many refills may be supplied for a particular medication.
- *Warnings*, which are provided by the prescriber with the intention of emphasizing specific concerns.

For example: if we look at a prescription for Patricia Pearson (see below), we can see that it is for Lipitor (atorvastatin Ca) 20 mg tablets, and that the patient is to receive 30 of them with 2 refills. The instructions to the patient would be, "*Take 1 tablet by mouth daily.*"

Dr. John Schoulties, M.D.
1650 Metropolitan St, Pittsburgh, PA 15233
Tel: (412) 555-4000   Fax: (412) 555-4790

Name _Patricia Pearson_   Date _8-31-2013_

Address   Age   Wt/Ht

R *Lipitor 20 mg*
   Disp: #30
   S: i tab po qd

Refills _2_

John Schoulties   M.D.   M.D.

Product Selection Permitted   Dispense As Written

DEA No. _____

Prescription No.: 00000212

Other things of note include the date that the prescription is written for is August 31, 2013. Prescriptions for non-controlled substances are only good for one year, so Mrs. Pearson will need a new script if she still needs this medication past August 31, 2014, regardless of how many refills were written for. Another noteworthy item is that the physician signed permitting product selection (i.e., generic substitution). The last significant item on this label is that the physician did not include their DEA number. A DEA number should only be used for controlled substances.

Besides over the counter medications (OTC) such as aspirin and ibuprofen, behind the counter medications (BTC) such as Allegra-D (fexofenadine with pseudoephedrine), and prescription medications (Rx legend) such as amoxicillin and digoxin, there is another group of medications to be concerned with called controlled substances. Controlled substances are medications with further restrictions due to abuse potential. There are 5 schedules of controlled substances with various prescribing guidelines based on abuse potential, as determined by the Drug Enforcement Administration and individual state legislative branches. CI medications are not available via a

245

prescription. CII medications may be written for a maximum 90 day supply, excluding hospice patients. No refills are allowed on schedule II medications. CIII-IV medications may only be written for a 6 month supply. CV medications may be written for up to 1 year, although many states limit this to 6 months.

Calculations

Various calculations involved in correctly interpreting prescriptions frequently need to be performed to account for days' supply, adjusting refills and short-fills. Prescribers often just include a quantity of medication to dispense and directions on how frequently to use it. They usually don't include the actual intended time frame. When prescribers write these prescriptions they may have various time frames in mind, which sometimes differ from the period of time required by a specific prescription benefits manager.

Days' supply

Physicians often just include a quantity of medication to dispense and directions on how frequently to use it. They usually don't include the actual intended time frame. As a pharmacy technician, you will need to translate that into Days' Supply.

In this chapter, Days' Supply is referring to how long a prescription order will last. Often, it is not as simple as giving a tablet once a day for 30 days; you will frequently need to do calculations for oral liquid medications, injectables, nasal sprays, and inhalers, and make estimations for PRN's, ointments and creams, lotions, eye and ear drops, and ophthalmic ointments.

Days' supply for tablets, capsules, and liquid medications

The first things to look at are tablets, capsules, and liquid medications because they're the most common and the most straight-forward to perform calculations with. Without wanting to over explain this process, let's look at some example problems.

Example: A prescription is written for amoxicillin 250 mg capsules #30 i cap t.i.d. What is the days' supply?

$$\frac{30\,caps}{1} \times \frac{1\,dose}{1\,cap} \times \frac{day}{3\,doses} = 10\,days$$

Example: A prescription is written for amoxicillin 250 mg/5 mL 150 mL i tsp tid. What is the days' supply?

$$\frac{150\,mL}{1}\times\frac{tsp}{5\,mL}\times\frac{dose}{1\,tsp}\times\frac{day}{3\,doses}=10\,days$$

The item to be careful about when it comes to tablets, capsules, and liquid medications are PRN medications, especially ones with variable doses and variable frequencies. In general, you should perform the calculations using the shortest interval with the highest dose. This will provide the shortest span of time in which they could use all the medication dispensed. Let's look at an example problem.

Example: A prescription is written for Ultram 50 mg #60 i-ii tabs po q4-6h prn pain. What is the days' supply?

$$\frac{60\,tabs}{1}\times\frac{1\,dose}{2\,tabs}\times\frac{4\,hours}{1\,dose}\times\frac{1\,day}{24\,hours}=5\,days$$

Conveniently, this example came out to an even number of days. Sometimes, your calculations will come out to a decimal number of days, and you may need to use some professional judgment to determine whether to drop the decimal or round up. If you are not sure, it is usually better to drop the decimal.

Days' supply for insulins

Most insulins are called U-100 insulins, meaning that each mL contains 100 units. Also, most insulin are available as either 10 mL vials or boxes of 5 syringes containing 3 mL in each syringe for a total of 15 mL in a box. A 10 mL vial of U-100 strength insulin would contain 1000 units, and a 15 mL box of syringes with U-100 insulin would contain 1500 units. This is good information to help you make quick work of the vast majority of insulin calculations. With insulin problems, whenever you come out with a decimal number of days you should always just drop the decimal, as you never want a diabetic patient to run out of their insulin. The last thing to keep in mind with respect to an insulin vial is that it should not be kept for longer than 30 days after it has been opened. Determining how long a box will last is different since each syringe is only good for 30 days after it is started, but there are five syringes. Let's look at an example problem with insulin.

Example: A prescription is written for Humulin N U-100 insulin 10 mL 35 units SC qd. What is the days' supply?

$$\frac{10\,mL}{1} \times \frac{100\,units}{mL} \times \frac{1\,day}{35\,units} = 28.57\,days$$

Which means **28 days**, because you should drop the decimal.

Days' Supply for inhalers and sprays

Whenever you see instructions on a product for a patient to receive a particular number of sprays or puffs of a given drug, you should stop and actually look at the packaging to discover how many metered inhalations or how many metered sprays are actually in the container. Lets look at an example problem accompanied by some of the text from the front of the container.

Example: A prescription is written for ProAir HFA 8.5 g inhaler 2 puffs q.i.d. What is the days' supply? (Additional information from medication package: 200 metered inhalations; 8.5 g net contents)

$$\frac{200\,puffs}{1} \times \frac{1\,dose}{2\,puffs} \times \frac{day}{4\,doses} = 25\,days$$

Days' supply for ointments and creams

Calculations for creams and ointments are a little more tricky because you usually don't know exactly how much will be used in a dose. The amount will depend on how large of an area is affected, and how many areas it needs to be applied to. The amount applied usually does not exceed 500 mg to 1 g, so unless you know otherwise, use 1 gram for the dose for each affected area.

Example: A prescription is written for Mycolog II cream 15 g apply sparingly bid. What is the days' supply?

$$\frac{15g}{1} \times \frac{1000\,mg}{1\,g} \times \frac{dose}{500mg} \times \frac{1\,day}{2\,doses} = 15\,days$$

Only 500 mg per dose is used in the calculation because the prescription specifically requested for it to be applied sparingly.

Days' supply for ophthalmic and otic preparations

To solve this type of problem, you need to know a conversion factor from milliliters to drops. The USP in chapter 1101 has written guidelines on the standardization of medicine droppers. Unless your specific medication notes something different, a dropper should be calibrated by the manufacturer to deliver between 18 and 22 drops per milliliter. Most people just split the difference and estimate **20 gtt/mL**. With that in mind, we can get a good estimate on how long the medication should last.

Another odd thing to keep track of when dealing with eye preparations are ophthalmic ointments. An ophthalmic ointment is typically applied as a very thin strip. Treat each dose of an ophthalmic ointment as 100 mg. Let's look at some example problems with respect to ophthalmic and otic preparations.

Examples: A prescription is written for timolol 0.25% Ophth. Sol. 5 mL i gtt ou q.d. What is the days' supply?

$$\frac{5\,mL}{1} \times \frac{20\,gtt}{1\,mL} \times \frac{1\,dose}{2\,gtt} \times \frac{1\,day}{1\,dose} = \textbf{50 days}$$

Example: A prescription is written for Neosporin Ophth. ung 3.5 g apply a thin strip ou q3-4h wa. What is the days' supply?

$$\frac{3.5\,g}{1} \times \frac{1000\,mg}{1\,g} \times \frac{1\,dose}{200\,mg} \times \frac{1\,day}{5\,dose} = \textbf{3 days } \textit{after you drop the decimal}$$

It is noteworthy that the request was to only apply the medication while awake. The standard expectation is for the patient to sleep 8 hours per day and therefore be awake for 16 hours a day. Under this expectation, the patient will likely receive 5 doses each day.

Days' supply for various packs

Many medications (such as birth control, steroids, and antibiotics) may come in packs with very explicit instructions for use. These instructions often explain exactly how many days they will last and require no additional calculations. You must simply read the instructions on the package to know how long it will last. Let's look at an example problem.

Example: A prescription is written for methylprednisolone 4 mg tabs taper pack use as directed. What is the days' supply?

The package has the following dosing information written on it:
1st Day: Take 2 tablets before breakfast, 1 tablet after lunch and

after supper, and 2 tablets at bedtime.

2nd Day: Take 1 tablet before breakfast, 1 tablet after lunch and after supper, and 2 tablets at bedtime.

3rd Day: Take 1 tablet before breakfast and 1 tablet after lunch, after supper, and at bedtime.

4th Day: Take 1 tablet before breakfast, after lunch, and at bedtime.

5th Day: Take 1 tablet before breakfast and at bedtime.

6th Day: Take 1 tablet before breakfast.

It requires nothing more than observation to realize it will last 6 days.

Days' supply for other miscellaneous medications

Unfortunately, there are many medications that have their own specific rules, such as estrogens given for hormone replacement therapy that are cycled on and off. The cycle is days 1-25 on, followed by five days off; or it may be cycled for three weeks on, followed by one week off. Items like a vial of nitroglycerin sublingual tablets or spray are expected to last a patient 30 days. With items like vaginal preparations, you'll need to know how much is delivered via the applicator. Lotions can be a challenge depending on their viscosity. A good rule of thumb for a lotion is to expect 2 mL to be used on each affected area per application. As you can tell, there are many individual little rules to try and work with when estimating how long a particular medication will last.

Adjusting refills and short-fills

The amount of medication a pharmacy can dispense to a patient is restricted first, by the prescriber's guidelines and second, by the insurer's guidelines. As previously mentioned in this chapter, when prescribers write these prescriptions, they may have various time frames in mind which sometimes differ from the period of time required by a specific prescription benefits manager. Calculations are often needed to first adjust the quantity dispensed to comply with the insurer's guidelines, and then the number of refills allowed.

To avoid over explaining this, let's look at a couple of example problems.

Example: A prescription is written for Dyazide #50 i cap p.o. q.d. + 3 refills. The insurance plan has a 30 day supply limitation. How many capsules can be dispensed using the insurance plan guidelines, and how many refills are allowed with the adjusted quantity?

First, calculate the total number of capsules allowed by the prescriber:

$$50\,capsules \times 4\,total\,fills\,(original\,fill + 3\,refills) = 200\,capsules$$

Then, using dimensional analysis, we can figure out how many capsules will be needed for each fill.

$$\frac{1\,cap}{dose} \times \frac{1\,dose}{day} \times \frac{30\,days}{fill} = \frac{30\,caps}{fill}$$

Next, using dimensional analysis, we can figure out how many fills from the insurer will be required to dispense the quantity written for by the physician.

$$\frac{200\,caps}{1} \times \frac{fill}{30\,caps} = 6.666\ldots fills$$

Therefore, there will be **5 refills** after the initial fill is dispensed, but there is still a partial fill left.

Now, based on the information we have already determined, we will figure out how many capsules to dispense for the partial fill.

$$200\,capsules - \left(\frac{6\,fills}{1} \times \frac{30\,caps}{fill} \right) = 20\,capsules$$

Now, let's restate everything in short:
We are dispensing **30 caps for our initial fill**.
The patient can have **5 refills** of 30 caps,
and a **partial fill of 20 caps**.

Quick review question 15

A prescription is written for NTG 2% ung, Disp 48 g, Sig: apply 1" qam and 6 hours later, Refills: 11. What is the dosage form of the medication?

a. sublingual spray
b. injectable solution
c. gel
d. ointment

Quick review question 16

A prescription is written for Timoptic 0.5%, Disp: 15 mL, Sig: i gtt o.d. bid, Refills: 2. What is the route of administration in this prescription?

a. left ear
b. right ear
c. left eye
d. right eye

Quick review question 17

A prescription is written for Humulin R cartridges, Disp: 1 box, Sig: 12 units SQ tid a.c., Refills: 5. When should this medication be given?

a. in the morning
b. before meals
c. after meals
d. twice a day

Quick review question 18

A prescription is written for ondansetron 4 mg, Disp: 9 tabs, Sig: i tab po q8h prn n/v, Refills: NR. What condition is this prescription attempting to ameliorate?

a. cholestatic pruritis
b. by mouth
c. nausea and vomiting
d. no refills

Quick review question 19

A prescription is written for E.E.S. 200 mg/5 mL, Disp 200 mL, Sig: iss tsp po tid ac for 7 days, Refills: NR. How much medication should the patient take for one dose?

a. one-half teaspoonful
b. one tablespoonful
c. one and one-half tablespoonful
d. one and one-half teaspoonful

Quick review question 20

A prescription is written for Benadryl cm, Disp 1 oz, Sig: aa 3 to 4 times daily prn itchiness. In the context of this prescription, what does the abbreviation 'aa' mean?

a. apply
b. of each
c. affected area
d. before meals

Quick review question 21

A prescription is written for prochlorperazine 25 mg, Disp #12, Sig: i supp pr q12h prn n/v, Refills NR. What is the dosage form of this medication?

a. suppository
b. tablet
c. injectable solution
d. none of the above

Quick review question 22

A prescription is written for simvastatin 20 mg, Disp: #90, Sig: i tab po q.d., Refills: 3. How often is the medication being given?

a. every other day
b. daily
c. three times a day
d. four times a day

Quick review question 23

A prescription is written for Zmax 2 G, Disp: 60 mL, Sig: Take entire contents immediately after reconstitution, Refills: NR. How much medication is the patient receiving in one dose?

a. 2 grains
b. 1 fluid ounce
c. 60 microliters
d. 2 grams

Quick review question 24

A prescription is written for prednisone 5 mg tabs, Disp: 1 taper pack, Sig: Take u.d., Refills: NR. In the context of this prescription, what does the abbreviation 'u.d.' mean?

a. unit dose

b. good directions
c. as directed
d. as needed

Quick review question 25

When should a prescriber include their DEA number on a prescription form?

a. only if the prescription is for a CII
b. on all orders for controlled substances
c. on all legend drugs
d. A physician should never include their DEA number on a prescription form.

Quick review question 26

What does it mean if a prescription order form has two signature lines that read "Product Selection Permitted" and "Dispense As Written", and the prescriber signs above the line that reads "Product Selection Permitted"?

a. The pharmacy may substitute a generic product if it is available on the market.
b. DAW 1
c. The pharmacy must dispense the brand the prescriber wrote for.
d. DAW 2

Quick review question 27

What does it mean if a prescription order form has two signature lines that read "Product Selection Permitted" and "Dispense As Written" and the prescriber signs above the line that reads "Dispense As Written"?

a. The pharmacy may substitute a generic product if it is available on the market.
b. DAW 0
c. The pharmacy must dispense the brand the prescriber wrote for.
d. DAW 2

Quick review question 28

A prescription is written for metformin 500 mg, Disp: 180, Sig: i tab po bid. How many days' supply does this prescription allow for?

a. 90 days
b. 30 days
c. 180 days
d. 28 days

Quick review question 29

A prescription is written for nizatidine 150 mg, Disp: 120, Sig: i cap bid ac. How many days' supply does this prescription allow for?

a. 30 days
b. 40 days
c. 60 days
d. 90 days

Quick review question 30

A prescription is written for Cortisporin Otic, Disp: 10 mL, Sig: iv gtt au qid. How many days' supply does this prescription allow for?

a. 6 days
b. 12 days
c. 13 days
d. 25 days

Quick review question 31

A prescription is written for Lantus insulin 10 mL, Disp: 1 vial, Sig: 30 U SQ qd. How many days' supply does this prescription allow for?

a. 33 days
b. 30 days
c. 28 days
d. 3 days

Quick review question 32

A prescription is written for Crestor 10 mg, Disp: #30, Sig: i tab po qd, Refills: 11. How should this prescription be adjusted if the patient prefers a 90 day supply in order to get three months worth of medication for only a two month supply copay (a savings offered through his insurance)?

a. This prescription cannot be adjusted.
b. dispense 90 tablets per fill with 11 refills and no partial fills

c. dispense 90 tablets per fill with 2 refills and a partial fill of 60 tablets

d. dispense 90 tablets per fill with 3 refills and no partial fills

Quick review question 33

A prescription is written for metoprolol tartrate 25 mg tabs, Disp: #100, Sig: ss tab po bid, Refills: 2. How should you adjust this if the patient were going to receive a thirty day supply for each filling of this medication?

a. dispense 30 tablets per fill with 9 refills and no partial fills

b. dispense 60 tablets per fill with 5 refills and no partial fills

c. dispense 30 tablets per fill with 5 refills and a partial fill of 20 tablets

d. dispense 60 tablets per fill with 2 refills and a partial fill of 20 tablets

Quick review question 34

A prescription is written for Glucotrol 10 mg tabs, Disp: #120, Sig: 20 mg po bid, Refills 5. How should this prescription be adjusted if the patient prefers a 90 day supply in order to get three months worth of medication for only a two month supply copay (a savings offered through his insurance)?

a. dispense 180 tablets per fill with 3 refills and no partial fills

b. dispense 90 tablets per fill with 7 refills and no partial fills

c. dispense 360 tablets per fill with 1 refill and no partial fills

d. dispense 90 tablets per fill with 5 refills and no partial fills

Quick review question 35

A prescription is written for Adderall XR 25 mg, Disp: #90, Sig: 1 cap po qam, Refills: NR. The patient's insurance has a thirty day dispensing limit on Schedule II medications. How should this prescription be adjusted to account for all 90 capsules that the physician wrote for?

a. dispense 30 tablets per fill with 2 refills and no partial fills

b. dispense 30 tablets per fill with no refills and no partial fills

c. The insurance cannot limit a prescription.

d. This prescription cannot be adequately adjusted because Schedule II medications are not allowed to receive refills.

Order entry

Once the information on a particular prescription has been accurately interpreted, it will need to be entered into the pharmacy software management system. To cover this, we should look at what pharmacy software management systems entail, and what kind of information will need to be entered to process the prescription.

Pharmacy management software systems

In order to process the prescription, it will need to be entered into a pharmacy management software system. Pharmacy management software systems are not standardized in appearance, which allows companies to compete with each other to try and implement their own vision as to how an integrated environment should look and feel for the end users. This translates into a plethora of competing systems on the market. You'll find that most pharmacies have chosen different systems, but no matter what they look like, they need to manipulate the same data.

When comparing systems, the first thing you'll notice about pharmacy software is that some programs have chosen very different user interface paradigms. For most programs, there are two types of interfaces: text-based user interfaces (somewhat reminiscent of DOS style programs in appearance) or graphical user interfaces (GUI, pronounced goo-ey).

Text-based user interfaces have the advantages that they can usually run on older hardware successfully and don't require as much network bandwidth. These types of interfaces tend to be faster and rely primarily on keyboard input. Most of these programs have been around a long time so they tend to be very mature and have relatively few software bugs. The down side is that newer employees tend to be intimidated by text-based user interfaces as they have grown accustomed to GUI interfaces in most of their interactions with software.

GUIs have the advantage that they are primarily point and click software relying largely on mouse usage along with keyboard input for text boxes (some GUIs are looking at possible transitions to incorporating touch interfaces). This kind of software may be installed locally or remotely, depending on the offerings from the manufacturer and softwares specific design. These systems will typically require more overhead when compared to text-based user interfaces, but staff often finds them less intimidating.

Some companies attempt to create software that can bridge the gap between these two paradigms. Speed Script, pictured below, combines elements of both.

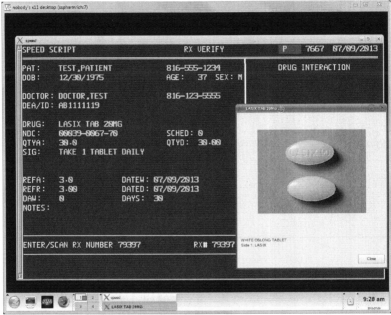

Most pharmacy management systems are capable of performing the following tasks:

- prescription processing
- e-prescribing
- prescription pick-up with signature capture
- refill queue
- work flow manager
- prescription scanning and patient card scanning
- med guides, care points, and REMS

- NDC check and prescription verification processing
- third party adjudication, including major medical insurance and worker's comp billing
- accounts receivable processing
- point of sale capabilities
- DME/CMS 1500 on-line billing
- 340B inventory control management and reports
- wholesaler interface with automated ordering
- nursing home and LTC batch processing
- wireless signature capture/delivery manifest module
- MTM services billing electronically
- compounding services support

A major challenge to any pharmacy when choosing a pharmacy management system is pricing along with per-seat licensing; types of tech support and maintenance; local servers versus remote hosting; accessing systems with a local client, compared to a web browser, compared to virtualization; the types of back-up services available; potential concerns over vendor lock-in, etc.

Data entry

Pharmacy staff will need to enter information into the pharmacy software management system in order to process the prescription. Common information to require includes:

- *Prescriber information* -This typically includes the prescriber's name, address of practice, contact information, medical license number, DEA number, and National Provider Identifier (NPI).
- *Third-party payor* - This includes coverage type (primary, secondary, etc.), insurance name and bank identification number (BIN), group number, and member number.
- *Patient information* - The patient information should at least include name, date of birth, address, contact information, allergies, and payment type (cash vs. insurance). Often, pharmacies will request information on concurrent use of other medications and dietary supplements, preferences with respect to safety lids, and verification that the patient

has received notification of the pharmacy's privacy policy.

- *Prescription information* - While many items on a prescription are important, the system should record as a minimum the date the prescription was written, superscription, inscription, subscription, signatura, refills, prescription origin code, and it should generate a unique prescription number that should appear on the prescription label as well.
- *DAW codes* - Dispense as written (DAW) codes need to be entered into the computer as well. Most prescriptions allow for generic substitution, and patients are glad to receive the more affordable version; therefore, the default DAW code is typically set to '0'. If a physician requires a specific medication to be dispensed, they will typically note this on the prescription. This is considered a DAW code of '1'. Sometimes a patient may request that they receive a brand name product even if a prescriber allowed for generic substitution. This would be classified as a DAW code of '2'. Other DAW codes are less frequently used. The following is a succinct list of the other DAW codes: 3 = substitution allowed - pharmacist selected product dispensed; 4 = substitution allowed - generic drug not in stock; 5 = substitution allowed - brand drug dispensed as generic; 6 = override; 7 = substitution not allowed - brand drug mandated by law; 8 = substitution allowed - generic drug not available in marketplace; and 9 = other.
- *Drug information* - At a minimum, drug information should include the drug name, the medication's National Drug Code (NDC), the manufacturer, and an ability to check for interactions and contraindications. Often, this drug information will include information on auxiliary labels, specific lot numbers and expiration dates, stock availability, pricing, and medication guides.

Quick review question 36

Which of the following statements about pharmacy management software systems is false?

a. There are a plethora of competing pharmacy management software systems on the market.

b. The interfaces for pharmacy management software systems have been standardized.

c. Despite the many programs on the market, there are some common information requirements that all pharmacies will need to input to process prescriptions.

d. The price structure and level of tech support can vary a great deal between various pharmacy management software systems.

Quick review question 37

What are some common items you will want to enter into the computer with respect to prescribers?

a. DEA number
b. medical license
c. NPI
d. all of the above

Quick review question 38

What is the purpose of a DAW code?

a. DAW codes are used to provide a quick explanation of whether or not a generic version of the medication is allowed to be dispensed, and if not, then why and who deemed the brand name product to be necessary.

b. DAW codes are mandated by HIPAA.

c. DAW codes were created to make the life of a pharmacy professional more complicated.

d. DAW codes do not fulfill a purpose.

Quick review question 39

Which DAW code should be used if a prescriber allows for generic substitution, but the patient insists that they receive the brand name product?

a. DAW 0
b. DAW 1
c. DAW 2
d. DAW 3

Quick review question 40

What kind of user interface do pharmacy management software

systems employ?

a. graphical user interfaces (GUIs)
b. text-based user interfaces
c. a combination of GUI and text-based user interface
d. all of the above are viable options for user interfaces

Quick review question 41

Which of the following items are pertinent patient information for pharmacy management software systems to track?

a. medication allergies
b. safety lid preferences
c. notification of privacy policies
d. all of the above

Quick review question 42

Which of the following items found on a prescription form would not be likely to get entered into the pharmacy management software system?

a. superscription
b. inscription
c. subscription
d. signatura

Filling the prescription

After completing the order entry process, the next step is to fill the medication order. A label will have been generated. You will want to verify that the information on the label is correct and complete. Then you will fill it with the appropriate medication, package it properly, and validate that everything has been done correctly.

Label

The minimal information which must appear on a dispensed prescription label includes:

- name and address of the pharmacy,

- serial (prescription) number,
- date of its filling,
- the name of the prescriber,
- the name of the patient,
- the directions for use, and
- any applicable warning (auxiliary) labels.

Also, on the federal level, the Comprehensive Drug Abuse Prevention and Control Act, also known as the Controlled Substances Act (CSA), requires on controlled substances the statement: *"Caution: Federal law prohibits the transfer of this drug to any person other than the patient for whom it was prescribed."*

Even though it is not required on the federal level, most prescription labels will also include the medication name, strength, and dosage form. Various states may provide additional guidelines for prescription labeling.

The following is an example of a dispensed prescription label.

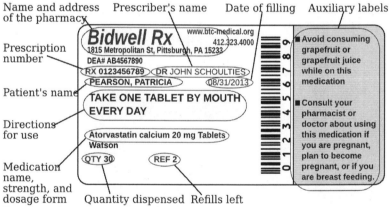

Name and address of the pharmacy   Prescriber's name   Date of filling   Auxiliary labels

Prescription number

Patient's name

Directions for use

Medication name, strength, and dosage form

Quantity dispensed   Refills left

Packaging

When it comes to packaging a product before it leaves the pharmacy, the prescription label is only one part of that process. The packaging must be suitable for maintaining the stability of the product and offering suitable safety to both the patient and others that may have access to the medication (i.e., children). With this in mind, we need to consider the following items: whether or not the packaging provides adequate light resistance; proper temperature for long term and short term storage; as some products will react to some plastics you need to consider PVC vs glass containers; child safety lids are concern to protect unintentional poisonings, but this needs to be balanced by sufficient ease of use for the patient; some products will require syringes (either oral or injectable), as a degree of knowledge about the intended dosage and dosage route of the medication will help with this decision.

Child safety caps

As a result of the Poison Prevention Packaging Act (PPPA), all prescription drugs and controlled dangerous substances must be packaged in child-resistant containers (i.e., packages with safety caps), with limited exceptions. The most common exceptions are if the patient or physician requests that the patient not receive child safety caps, or for emergency sublingual medications commonly used in response to angina.

If a medication is dispensed without a child safety cap, most pharmacies will note it in their pharmacy management software and often include an explanation as to why. Failure to comply with the PPPA could result in the pharmacist being prosecuted and imprisoned for not more than 1 year, or sentenced to pay a fine of not more than $1000, or both.

Light resistance

There are more than 200 different medications which are light sensitive. The chemical composition of these medications can be altered by exposure to direct light. As an example, when nitroprusside is exposed to direct sun light, it will breakdown into cyanide. Some common light sensitive medications include acetazolamide, doxycycline, linezolid, and zolmitriptan. While many drugs need to be protected from light while in storage, their original package from the manufacturer should suffice. If you need to repackage any medications, always be sure to consult the manufacturer's recommendations to determine if you need to place the medication in light resistant packaging or not.

Temperature

The majority of medications are safe to store at controlled room temperature, 15° to 30° C or 59° to 86° F. Some medications may require refrigeration, 2° to 8° C or 36° to 46° F, or they may even need to be frozen, -25° to -10° C or -13° to 14° F, after they are dispensed (these temperature ranges have been established by the United States Pharmacopeia Convention). If the patient is picking them up from a community pharmacy, the pharmacy staff should inform the patient if special storage requirements are needed. If these medications are being dispensed by either a delivery service or a mail order pharmacy, extra steps may need to be taken to help ensure that the products are maintained within proper temperature ranges, such as using cold containers with ice packs.

Container materials

Various plastics, including polyvinyl chloride (PVC), have been used in heath care for over 50 years, as it is an affordable and tear resistant product that can be manipulated to achieve varying degrees of flexibility versus rigidity. For most products, this is an ideal packaging material; however, some medications either contain or function as plasticizers, which cause them to bind to the PVC and also change the fluidity of the PVC container itself. These medications need to be stored or manipulated with products that they will not react to. Glass is a traditional option as it is an inert substance. In situations where plastics are still desirable, the plastics will need to be manipulated to prevent the medications from functioning as a plasticizer by either coating the surfaces of the container or a chemical change throughout the whole structure of the container. Nitroglycerin is a common example of this where the sublingual form and injectable forms both need to avoid prolonged exposure to traditional PVC containers.

Syringes

For patient safety, there needs to be a distinction between when to use oral/topical syringes and injectable syringes. If an oral or topical medication needs to be precisely dosed, an oral/topical syringe may be an ideal tool to dispense for the patient. They are simple to use, and provide a safety factor that needles cannot easily be attached to them ensuring that the patient does not receive them as an injection.

Injectable syringes also come in a wide variety of sizes with various needle options. Traditionally, prescribers will include an order for the patient's syringes and needles; although, the pharmacy staff may still need to determine the most appropriate syringes to dispense. In most states, patients may also request syringes without a prescription, but they may have difficulty getting a third-party payor to cover them.

Product validation

There are several means that may be utilized to provide proper product validation, including the use of NDCs, barcode scanning technology, and a final visual verification. The pharmacy management software will have recorded a specific NDC for the medication that is being dispensed. It is a good practice to check the NDC in the computer against the product being dispensed to ensure both accuracy and to avoid any kind of billing fraud. Many pharmacy management systems also include barcode scanning to provide another double check that the NDC on the bottle matches the NDC in the computer (the barcode on the manufacturer's bottle contains the NDC). The oldest and most common form of verification is visual inspection. Traditionally, the pharmacist provides this role, but in some states pharmacy technicians, under specific conditions, are allowed to perform tech check tech. This visual verification should be looking for an accurate interpretation of the information from the original prescription to the prescription label, that the patient information on the product is correct, that the product is packaged in an appropriate manner, and that the correct drug is being dispensed.

Quick review question 43

According to federal law, what statement must appear on all controlled substances?

   a. This medication may be habit forming.
   b. Do not attempt to drive a car or operate heavy machinery while under the influence of this medication.
   c. Caution: Federal law prohibits the transfer of this drug to any person other than the patient for whom it was prescribed.
   d. There are no statements required by federal law that have to appear on all controlled substances.

Quick review question 44

The majority of medications need to be stored at controlled room temperature. What is controlled room temperature according to the USP?

a. -25° to -10° C
b. 2° to 8° C
c. 15° to 30° C
d. 36° to 46° C

Quick review question 45

Which law put in place the regulations relating to child safety caps?

a. Poison Prevention Packaging Act
b. Occupation Safety and Health Act
c. Hazardous Drug Amendment
d. Federal Anti-Tampering Act

Quick review question 46

Nitroglycerin sublingual tablets should be stored in what kind of container?

a. PVC
b. glass vial
c. directly in a traditional amber vial with a child safety cap
d. none of the above

Quick review question 47

What information is present in a manufacturer's barcode on a prescription vial?

a. NDC
b. lot number
c. expiration date
d. all of the above

Quick review question 48

Which of the following items does not need to appear on a dispensed prescription vial?

a. patient name
b. patient address
c. pharmacy name

d. directions for use

Quick review question 49

Why should oral/topical medications not be dispensed in injectable syringes?

a. injectable syringes are more expensive
b. to prevent accidental injection
c. the measurements on an oral/topical syringe are more precise
d. none of the above

Medication pickup

Once the medication has been filled and checked, it is ready for the patient to pick it up, which usually involves any payments (including copays), the offering of medication counseling, and the proper recording and filing of dispensed prescriptions.

In a community pharmacy setting, a patient (or their family member, caregiver, or other authorized individual) may go to the prescription pick-up counter to receive their medication. As some patients may have the same or similar names, many pharmacies require the answer to a simple question to ensure that the medication is for the individual at the counter. Common questions include asking for their birthday or the last four digits of their social security number, which the technician can verify in the pharmacy software. At this point, the technician needs to collect any necessary payments. Conveniently, many pharmacy software management systems are able to offer point of sale capabilities as well. The technician also needs to offer medication counseling to the patient, and the patient will then record in writing (whether on a physical piece of paper, or through electronic signature capture) if they want to consult the pharmacist about their therapy. If the patient requests medication counseling, they will need the pharmacist. Under the privacy provisions of HIPAA, the pharmacy needs to make a valid attempt at privacy during counseling whether with something as simple as a privacy screen, or with an entirely separate room. If the patient seems confused about their medication, it is a common practice for the technician to strongly suggest that the patient receive counseling or even just have the pharmacist assist the patient.

Pharmacies need to keep track of what medications have been dispensed through either the computer software and/or through the proper filing of hard copies. Prescription filing is a necessary part of working in a pharmacy. Title 21 of the Code of Federal Regulations - Section 1304.04 Maintenance of records and inventories - Subsection h, you will find the prescription filing options that are considered acceptable on the federal level (some states may place additional requirements on this process).

h) Each registered pharmacy shall maintain the inventories and records of controlled substances as follows:

(1) Inventories and records of all controlled substances listed in Schedule I and II shall be maintained separately from all other records of the pharmacy.

(2) Paper prescriptions for Schedule II controlled substances shall be maintained at the registered location in a separate prescription file.

(3) Inventories and records of Schedules III, IV, and V controlled substances shall be maintained either separately from all other records of the pharmacy or in such form that the information required is readily retrievable from ordinary business records of the pharmacy.

(4) Paper prescriptions for Schedules III, IV, and V controlled substances shall be maintained at the registered location either in a separate prescription file for Schedules III, IV, and V controlled substances only or in such form that they are readily retrievable from the other prescription records of the pharmacy. Prescriptions will be deemed readily retrievable if,

at the time they are initially filed, the face of the prescription is stamped in red ink in the lower right corner with the letter "C" no less than 1 inch high and filed either in the prescription file for controlled substances listed in Schedules I and II or in the usual consecutively numbered prescription file for noncontrolled substances. However, if a pharmacy employs a computer application for prescriptions that permits identification by prescription number and retrieval of original documents by prescriber name, patient's name, drug dispensed, and date filled, then the requirement to mark the hard copy prescription with a red "C" is waived.

(5) Records of electronic prescriptions for controlled substances shall be maintained in an application that meets the requirements of part 1311 of this chapter. The computers on which the records are maintained may be located at another location, but the records must be readily retrievable at the registered location if requested by the Administration or other law enforcement agent. The electronic application must be capable of printing out or transferring the records in a format that is readily understandable to an Administration or other law enforcement agent at the registered location. Electronic copies of prescription records must be sortable by prescriber name, patient name, drug dispensed, and date filled.

You may view this on the DEA's Diversion Control website at

Also, while not a requirement, a common pharmacy practice is to stamp their Schedule-II medication with an 'N' stamp to denote it as a narcotic and make it easily identifiable from other prescriptions.

Quick review question 50

Why do pharmacies often ask patient verification questions?

a. to prevent dispensing medication to the incorrect recipient
b. to annoy customers
c. to comply with TJC requirements
d. to comply with HIPAA regulations

Quick review question 51

Who may provide patient counseling?

a. any member of the pharmacy staff
b. the pharmacist(s)
c. pharmacy technicians may if they are in a state that allows tech-check-tech
d. answers b and c are both correct

Quick review question 52

If a pharmacy maintains hard copies of their prescriptions, which prescriptions should be marked with a red colored 'C'?

a. C II medications
b. C III, C IV, and C V medications
c. all controlled substances
d. all legend drugs

Quick review question 53

Which regulation requires pharmacies to make a reasonable attempt to provide privacy for patients receiving counseling?

a. OBRA-90
b. Kefauver Harris Amendment
c. HIPAA
d. CSA

Quick review question 54

When a patient is picking up their medication, and the patient is asked to sign, what are they signing for?

a. proof that they received their medication
b. whether or not they want child safety caps used on their prescriptions
c. whether or not they want medication counseling
d. Patients do not need to sign for anything when they pick-up their medications.

Quick review question 55

If a patient seems confused about their medications, what should a pharmacy technician do?

a. assure the patient that everything will be fine
b. tell the patient to call their physician
c. provide a sympathetic ear, but do nothing else
d. have the pharmacist assist the patient

Quick review question 56

Which of the following statements about prescription filing is true?

a. There are specific federal regulations with respect to how prescriptions should be filed.
b. Some states have their own guidelines/regulations with respect to prescription filing.
c. Some pharmacy management software systems can be used for prescription filing by storing images of the prescriptions and providing them in a readily retrievable format.
d. all of the above

Answers to quick review questions

1. D - E-prescribing is the computer-based electronic generation, transmission and filling of a medical prescription, taking the place of paper and faxed prescriptions. E-prescribing allows a physician, nurse practitioner, or physician assistant to electronically transmit a new prescription or renewal authorization to a community or mail-order pharmacy.

2. C - CPOE systems are used for processing orders in institutional settings (which includes hospitals).
3. C - The signatura (also called sig, or transcription), gives instructions on a prescription to the patient on how, how much, when, and how long the drug is to be taken.
4. A - A pharmacy benefits manager (PBM) is a company that acts as an intermediary between the pharmacy and the insurance plan.
5. B - On a health insurance card, a bank identification number (BIN) is a six digit number used to identify a specific plan from a carrier making it easier for the PBM to process your prescription online. No actual bank is involved in this part of the process; the name is a hold over from early electronic banking term.
6. A - A DAW code of '0' applies to most prescriptions as they allow for generic substitution, and patients are generally willing to receive the more affordable version.
7. B - If a physician requires a specific medication to be dispensed, they will typically note this on the prescription. This is considered a DAW code of '1'.
8. D - Prescriptions may arrive on a traditional prescription form (there are specific security requirements on this if the patient is using Medicaid), a fax or phone call from an appropriately licensed prescriber, e-prescribing in an outpatient setting, and computerized prescriber order entry (CPOE) in an institutional setting.
9. A - If the order is for a controlled substance, the prescriber's DEA number must be included on the prescription form. Community pharmacies and mail order pharmacies will need NPIs and medical licenses for the various prescribers, but they do not need to be on the prescription form.
10. B - The Medicaid Tamper-Resistant Prescription Pad Law applies specifically to Medicaid patients.
11. A - E-prescriptions may be used for all patients.
12. C - Through the requirement of using computer programs that communicate through SureScripts to electronically prescribe outpatient prescriptions, SureScripts has the responsibility of authenticating both the receipt and delivery of the electronic prescription.
13. B - While CPOE provides a plethora of advantages (CPOE

decreases delay in order completion, reduces errors related to handwriting or transcription, allows order entry at the point of care or off-site, provides error-checking for duplicate or incorrect doses or tests, and simplifies inventory and posting of charges), implementation costs are not typically considered one of them.

14. D - When a patient first arrives at a community pharmacy, the pharmacy team will need to gather/verify various important pieces of patient data including gathering drug and disease information, ensuring that the pharmacy has the correct name, address, contact information, and any other pertinent data, documenting/updating allergy information, and verifying/updating medication insurance information.

15. D - The abbreviation 'ung' stands for ointment.

16. D - The abbreviation 'o.d.' stands for right eye.

17. B - The abbreviations 'tid ac' stands for three times a day before meals.

18. C - The prescription is intended to ameliorate, or improve, the patient's nausea and vomiting (n/v).

19. D - The abbreviations 'iss tsp' stands for one and one-half teaspoonful.

20. C - In the context of this prescription, the abbreviation 'aa' stands for affected area.

21. A - The abbreviation 'supp' stands for suppository.

22. B - The abbreviation 'q.d.' can be translated as every day or daily.

23. D - The dose is 2 grams since the entire bottle is being taken.

24. C - In the context of this prescription, the abbreviation 'u.d.' stands for as directed.

25. B - A DEA number should only be used for controlled substances.

26. A - By marking the prescription as "Product Selection Permitted", it allows for generic substitution.

27. C - By marking the prescription as "Dispense As Written", it is stating that the product listed on the prescription is the brand that must be dispensed.

28. A - The problem can be solved as follows:

$$\frac{180\,tabs}{1} \times \frac{1\,dose}{1\,tab} \times \frac{day}{2\,doses} = \mathbf{90\,days}$$

29.C - The problem can be solved as follows:

$$\frac{120\,caps}{1}\times\frac{1\,dose}{1\,cap}\times\frac{day}{2\,doses}=\mathbf{60\,days}$$

30.A - The problem can be solved as follows:

$$\frac{10\,mL}{1}\times\frac{20\,gtt}{1\,mL}\times\frac{1\,dose}{8\,drops}\times\frac{day}{4\,doses}=6.25\,days$$

after you drop the decimal it becomes **6 days**

31.B - The problem can be solved as follows:

$$\frac{10\,mL}{1}\times\frac{100\,units}{1\,mL}\times\frac{1\,dose}{30\,units}\times\frac{day}{1\,dose}=33.\overline{3}\,days$$

*Since an insulin vial is only good for 30 days after
the vial is pierced it becomes* **30 days**

32.D - The problem can be solved as follows:

$30\,tabs\times12\,total\,fills(\,original\,fill+11\,refills)=360\,tabs$

$$\frac{1\,tab}{dose}\times\frac{1\,dose}{day}\times\frac{90\,days}{fill}=90\,tabs\,per\,fill$$

$$\frac{360\,tabs}{1}\times\frac{fill}{90\,tabs}=4\,fills$$

Therefore, dispense **90 tablets per fill**
with **3 refills**
and **no partial fills**

33.A - The problem can be solved as follows:

$100\,tabs\times3\,total\,fills(\,original\,fill+2\,refills)=300\,tabs$

$$\frac{0.5\,tab}{dose}\times\frac{2\,doses}{day}\times\frac{30\,days}{fill}=30\,tabs\,per\,fill$$

$$\frac{300\,tabs}{1}\times\frac{fill}{30\,tabs}=10\,fills$$

Therefore, dispense **30 tablets per fill**
with **9 refills**
and **no partial fills**

34.C - The problem can be solved as follows:

$$120\,tabs \times 6\,total\,fills\,(original\,fill + 5\,refills) = 720\,tabs$$

$$\frac{2\,tab}{dose} \times \frac{2\,doses}{day} \times \frac{90\,days}{fill} = 360\,tabs\,per\,fill$$

$$\frac{720\,tabs}{1} \times \frac{fill}{360\,tabs} = 2\,fills$$

Therefore, dispense **360 tablets per fill**
with **1 refill**
and **no partial fills**

35. D - At the federal level, prescription for Schedule II medications are not allowed to receive refills, therefore this prescription can not be adequately adjusted if the insurance will only cover 30 days at a time. Either the patient will need to pay for the medication out of pocket, or get three prescriptions for the medication, each allowing for a thirty day supply.

36. B - Pharmacy management software systems are not standardized in appearance.

37. D - Prescriber information typically includes the prescriber's name, address of practice, contact information, medical license number, DEA number, and National Provider Identifier (NPI).

38. A - DAW codes are used to provide a quick explanation of whether or not a generic version of the medication is allowed to be dispensed, and if not then why and whom deemed the brand name product to be necessary.

39. C - Sometimes a patient may request that they receive a brand name product even if a prescriber allowed for generic substitution. This would be classified as a DAW code of '2'.

40. D - Pharmacy management software systems use a text-based user interfaces (somewhat reminiscent of DOS style programs in appearance), a GUI (pronounced goo-ey and stands for graphical user interface), or some companies create software that makes use of both types of interfaces.

41. D - The patient information tracked by pharmacy management software should at least include name, date of birth, address, contact information, allergies, and payment type (cash vs. insurance). Often, pharmacies will request

information on concurrent use of other medications and dietary supplements, preferences with respect to safety lids, and verification that the patient has received notification of the pharmacy's privacy policy, all of which are also typically tracked by software.

42. A - While the medication name and strength (inscription), the quantity to dispense (subscription), and the directions for the patient (signatura) are all going to be entered into the pharmacy management software system, the 'Rx' symbol (superscription) is not likely to be entered into the software.

43. C - On the federal level, the Comprehensive Drug Abuse Prevention and Control Act, also known as the Controlled Substances Act (CSA), requires that on controlled substances the statement: *"Caution: Federal law prohibits the transfer of this drug to any person other than the patient for whom it was prescribed."*

44. C - The majority of medications are safe to store at controlled room temperature, 15° to 30° C or 59° to 86° F. Some medications may require refrigeration, 2° to 8° C or 36° to 46° F, or they may even need to be frozen, -25° to -10° C or -13° to 14° F, after they are dispensed (these temperature ranges have been established by the United States Pharmacopeia Convention).

45. A - As a result of the Poison Prevention Packaging Act (PPPA), all prescription drugs and controlled dangerous substances must be packaged in child-resistant containers (i.e., packages with safety caps), with limited exceptions.

46. B - Sublingual nitroglycerin needs to avoid prolonged exposure to traditional PVC containers, including direct exposure to traditional amber prescription vials. Glass is a safe alternative as it is an inert substance.

47. A - The barcode on the manufacturer's bottle contains the NDC, but no other information is in the barcode currently.

48. B - The patient's address does not need to appear on a prescription label. The minimal information which must appear on a dispensed prescription label includes name and address of the pharmacy, serial (prescription) number, date of its filling, the name of the prescriber, the name of the patient, the directions for use, and any applicable warning (auxiliary) labels.

49. B - Oral/topical syringes provide a safety factor that needles cannot easily be attached to them, ensuring that the patient does not receive them as an injection.

50. A - As some patients may have the same or similar names, many pharmacies require the answer to a simple question to ensure that the medication is for the individual at the counter.

51. B - If the patient requests medication counseling, they will need the pharmacist.

52. B - Paper prescriptions for Schedules III, IV, and V controlled substances shall be maintained at the registered location either in a separate prescription file for Schedules III, IV, and V controlled substances only, or in such form that they are readily retrievable from the other prescription records of the pharmacy. Prescriptions will be deemed readily retrievable if, at the time they are initially filed, the face of the prescription is stamped in red ink in the lower right corner with the letter "C" no less than 1 inch high, and filed either in the prescription file for controlled substances listed in Schedules I and II, or in the usual consecutively numbered prescription file for noncontrolled substances.

53. C - Under the privacy provisions of HIPAA, the pharmacy needs to make a valid attempt at privacy during counseling, whether with something as simple as a privacy screen, or with an entirely separate room.

54. C - The technician needs to offer medication counseling to the patient, and the patient will then record in writing (whether on a physical piece of paper, or through electronic signature capture) if they want to consult the pharmacist about their therapy.

55. D - If the patient seems confused about their medication, it is a common practice for the technician to strongly suggest that the patient receive counseling or even just have a pharmacist assist the patient.

56. D - Title 21 of the Code of Federal Regulations - Section 1304.04 subsection h specifies information on how to file prescriptions, including allowing for the use of pharmacy software under specific guidelines. Also, some states have added additional prescription filing requirements.

CHAPTER 7 Pharmacy Inventory Management

Key concepts

This chapter will cover the following knowledge areas to prepare you for the *Pharmacy Technician Certification Exam*:

- Function and application of NDC, lot numbers and expiration dates
- Formulary or approved/preferred product list
- Ordering and receiving processes (e.g., maintain par levels, rotate stock)
- Storage requirements (e.g., refrigeration, freezer, warmer)
- Removal (e.g., recalls, returns, outdates, reverse distribution)

Terminology

To get started with, there is some terminology that should be defined.

inventory - Inventory is simply the entire stock on hand for sale at a given time.

perpetual inventory - Perpetual inventory is a system that maintains a continuous count of every item in inventory so that it always shows the stock on hand. Some pharmacies maintain perpetual inventories on all products, while others only do this with their schedule II medications.

reorder point - A reorder point, also called a *par level*, is a minimum stock level which triggers when a product reorder should be placed.

formulary - A formulary is a list of medications available for use within a health care system. There are two major types of formularies: open formularies and closed formularies.

open formulary - An open formulary implies that the pharmacy must stock, or have ready access to, all drugs that may be written by the prescribers in their practice area.

closed formulary - Under a closed formulary, the drug inventory is limited to a list of approved medications.

purchasing - Purchasing is the ordering of products for use or sale by the pharmacy, and is usually carried out by either an independent or group process.

direct purchasing - Direct purchasing entails ordering medications directly from the original drug manufacturer. This typically requires completion of a purchase order - generally, a preprinted form with a unique number, on which the product name(s), amount(s), and price(s) are entered.

wholesaler purchasing - Wholesaler purchasing enables the pharmacy to use a single source to purchase numerous products from numerous manufacturers. Most drug ordering of this fashion is done online.

prime vendor purchasing - Prime vendor purchasing involves an agreement made by a pharmacy for a specified percentage or dollar volume of purchases in exchange for being given lower acquisition costs.

controlled substances - Controlled substances are medications with restrictions due to abuse potential. There are 5 schedules of controlled substances with various prescribing guidelines based on abuse potential, counter balanced by potential medicinal benefit as determined by the Drug Enforcement Administration and individual state legislative branches.

schedule II medications - Schedule II medications must be stocked separately in a secure place or distributed throughout your inventory, and require a DEA 222 form for reordering, whether using a triplicate paper version or an authorized electronic system. Their stock must be continually monitored and documented. Some states may place additional restriction on purchasing and storage requirements for these medications.

Occupational Safety and Health Administration (OSHA) - OSHA is a government agency within the United States Department of

Labor responsible for maintaining safe and healthy work environments.

Material Safety Data Sheet (MSDS) - OSHA-required notices on hazardous substances which provide hazard, handling, clean-up, and first aid information.

The Joint Commission (TJC) - The Joint Commission, formerly the Joint Commission on Accreditation of Healthcare Organizations (JCAHO), is a nonprofit organization that accredits more than 20,000 healthcare organizations and programs in the United States.

hazardous drugs - Hazardous drugs are drugs that are known to cause genotoxicity, which is the ability to cause a change or mutation in genetic material; carcinogenicity, which is the ability to cause cancer in animal models, humans or both; teratogenicity, which is the ability to cause defects on fetal development or fetal malformation; and lastly, hazardous drugs are known to have the potential to cause fertility impairment, which is a major concern for most clinicians. These drugs can be classified as antineoplastics, cytotoxic agents, biologic agents, antiviral agents and immunosuppressive agents.

reverse distribution - Reverse distribution involves the returning of medications to specialized brokers for management, and may involve a monetary credit to the provider. Reasons for using reverse distribution may vary, but often include controlled substances, expired or short-dated medications, excessive stock, and improperly stored products.

National Drug Code (NDC) - The National Drug Code is a unique product identifier used for medications intended for human use. It is a unique 10-digit, 3-segment numeric identifier assigned to each medication. The first segment identifies the manufacturer. The second segment identifies a specific strength, dosage form, and formulation for a particular manufacturer. The third segment identifies package forms and sizes.

lot number - A lot number is an identification number assigned to a particular quantity (or lot) of material from a single manufacturer made in a specific batch. Sometimes also called a batch number or control number.

Quick review question 1

Which of the following statements best defines the term 'inventory'?

a. the cumulative value of all the merchandise available for sale
b. the entire stock on hand for sale at a given time
c. all the stock and fixtures within a facility
d. none of the above

Quick review question 2

What term below would be defined as a list of medications available for use within a health care system?

a. inventory
b. purchase order
c. formulary
d. wholesaler

Quick review question 3

What is the proper term for when someone purchases directly from the manufacturer?

a. direct purchasing
b. wholesaler purchasing
c. prime vendor purchasing
d. none of the above

Quick review question 4

Which government agency is responsible for maintaining a safe and healthy work environment?

a. JCAHO
b. FDA
c. TJC
d. OSHA

Quick review question 5

What information is included in a National Drug Code?

a. manufacturer
b. medication strength and dosage form
c. package form and size

d. all of the above

Quick review question 6

Which of the following is not a reason for classifying a medication as hazardous?

a. it causes photosensitivity
b. it can cause cancer
c. it can cause fetal defects
d. it can impair fertility

Inventory management processes

The primary purpose of inventory management is the timely purchase and receipt of pharmaceuticals, and to establish and maintain appropriate levels of materials in stock. Purchasing, receiving, and inventory should be as uncomplicated as possible so as not to disrupt or to interfere with the other activities of the pharmacy. Community pharmacies usually maintain an open formulary, which means they must stock, or have ready access to, all drugs that may be written by the physicians in their practice area. In the community pharmacy, most pharmaceutical products are purchased through a local wholesaler. Specialty pharmacies and hospital pharmacies use a closed formulary, which means the drug inventory is limited to a list of approved medications. To minimize the cost of doing business, inventory levels must be adequate, but not excessive, with a rapid turnover of drug stock on the shelf. The purchasing, receipt, and inventory of controlled-drug substances requires special procedures and record-keeping requirements.

Purchasing

Purchasing is the ordering of products for use or sale by the pharmacy, and is usually carried out by either an independent or group process. In independent purchasing, the pharmacist or technician deals directly with a drug wholesaler (or rarely, the pharmaceutical manufacturer) regarding matters such as price and terms. In group purchasing, a number of pharmacies work together to negotiate a discount for high-volume purchases and more favorable contractual terms. Several purchasing methods are used in pharmacy, including Direct Purchasing, Wholesaler Purchasing, and Prime Vendor Purchasing.

Direct purchasing entails ordering medications directly from the original drug manufacturer. This typically requires completion of a purchase order - generally, a preprinted form with a unique number on which the product name(s), amount(s), and price(s) are entered.

Advantages to direct purchasing:

- lower cost
- lack of add-on fees

Disadvantages to direct purchasing:

- a commitment of time, as it will take longer than other methods to receive drugs
- a commitment of staff, since there are multiple requisitions to be completed and mailed to multiple pharmaceutical companies

Wholesaler purchasing enables the pharmacy to use a single source to purchase numerous products from numerous manufacturers. Most drug ordering of this fashion is done on-line, although gathering information may be done in a number of different ways, such as writing items on a 'want book', walking the shelves and scanning items that need reordered into a portable bar code scanner, or many pharmacy management software programs will automatically populate a reorder list when the pharmacy stock reaches a predetermined reorder point. A system that maintains a continuous inventory record is known as perpetual inventory.

Advantages to wholesaler purchasing:

- reduced turnaround time for orders
- lower inventory and associated costs
- reduced commitment of time and staff

Disadvantages to wholesaler purchasing:

- a higher purchase cost
- supply difficulties
- loss of control provided by in-house purchase orders

- unavailability of some pharmaceuticals

Sometimes, to offset some of the increased costs associated with using a wholesaler as opposed to direct purchasing, pharmacies will establish a contract identifying a particular wholesaler as a prime vendor purchaser. Prime vendor purchasing involves an agreement made by a pharmacy for a specified percentage or dollar volume of purchases in exchange for being given lower acquisition costs.

Advantages to prime vendor purchasing:

- lower acquisition costs
- competitive service fees
- electronic order entry
- often, emergency delivery services
- promotes just in time (JIT) purchasing

Disadvantages to prime vendor purchasing

- limits ability to use other wholesalers
- in terms of JIT, it can only be used when supplies are readily available and needs can be accurately predicted

Controlled substances, as expected, require special consideration when it comes to purchasing. The Controlled Substance Act (CSA) defines procedures for purchasing and receiving, and requirements for inventory and record keeping.

Schedule III – V drugs (see Chapter 2 for information on drug schedules) may be ordered by a pharmacy or other appropriate dispensary on a general order from a wholesaler, and you should check the delivery in against the original order.

Schedule II drugs have much more stringent requirements. A pharmacy must register with the Drug Enforcement Administration (DEA) to purchase Schedule II medications. The purchase of such controlled substances must be authorized by a pharmacist and executed on either a triplicate DEA 222 order form or an electronic 222 form through a controlled substances ordering system (CSOS).

The DEA form 222 is a triplicate form. The pharmacy retains the third sheet, while sending the first and second pages to the wholesaler. The wholesaler is responsible for sending the second page to the

DEA, while retaining the first page for its own records. When the Schedule II medications arrive in the pharmacy, they should be checked in against the DEA form.

Sample DEA Form 222								

See Reverse of PURCHASER's Copy for Instructions

No order form may be issued for Schedule I and II substance unless completed application form has been received. (21 CRF 1305.04)

OMB APPROVAL NO. 1117-0010

TO: STREET ADDRESS

CITY and STATE	DATE	TO BE FILLED IN BY SUPPLIER
		SUPPLIER DEA REGISTRATION No.

TO BE FILLED IN BY PURCHASER

	No. of Packages	Size of Package	Name of Item	National Drug Code	Packaging Shipped	Date Shipped
1						
2						
3						
4						
5						
6						
7						
8						
9						
10						

LAST LINE COMPLETED   *(MUST BE 10 OR LESS)*   SIGNATURE OF PURCHASER OR ATTORNEY OR AGENT

Date Issued	DEA Registration No	Name and Address of Registrant
Schedules		
Registered as a	No. of this Order Form	

DEA Form 222
(Oct. 1992)

US OFFICIAL ORDER FORMS - SCHEDULES I & II
DRUG ENFORCEMENT ADMINISTRATION
SUPPLIER'S Copy 1

286

The following is an image explaining how an electronic 222 form works using CSOS.

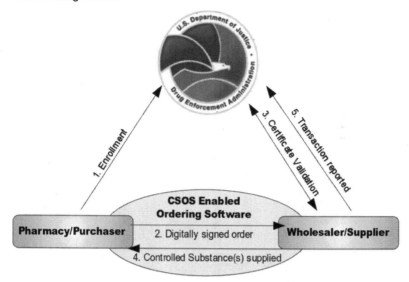

1. An individual enrolls with the DEA and, once approved, is issued a personal CSOS Certificate.
2. The purchaser creates an electronic 222 order using an approved ordering software. The order is digitally signed using the purchaser's personal CSOS Certificate, and then transmitted to the suppliers. The paper 222 is not required for electronic ordering.
3. The supplier receives the purchase order, and verifies that the purchaser's certificate is valid with the DEA. Additionally, the supplier validates the electronic order information just like it would a paper order.
4. The supplier completes the order and ships to the purchaser. Any communications regarding the order are sent electronically.
5. The order is reported by the supplier to the DEA within two business days.

Receiving

The pharmaceutical received should be carefully checked in against the purchase order, including product name, quantity, strength, and package size. Controlled substances are shipped in separate containers and should be checked in by a pharmacist (although pharmacy technicians may assist with this process under the direct supervision of a pharmacist). Schedule II medications need to be checked in against your DEA 222 form (whether the paper triplicate form, or the electronic form on your CSOS enabled software).

If something is damaged in shipment or improperly shipped, it must be reported to the pharmacist and vendor immediately.

Inventory

Inventory, simply put, is the entire stock on hand for sale at a given time. Inventory typically includes prescription drugs, over the counter medications, dietary supplements, and front end merchandise. Traditionally, inventory is the single largest expense in pharmacy, and so proper management of it is essential to the success of any pharmacy.

Most pharmacies must borrow and pay interest in order to keep an adequate quantity of medications on the shelves in order to meet anticipated needs. Most pharmacies establish maximum and minimum inventory levels to try and assure an adequate, but not excessive, quantity of medicines are on the shelf, so that they do not need to borrow too much money (or even if they aren't using loans, they don't want too much money tied up in non-moving inventory). Often, max and min levels are monitored by computer systems, but pharmacy staff will need to initially establish those levels, and may occasionally need to intervene and adjust those levels. Furthermore, some pharmacies still do not have automated inventory systems. When you establish a max level, that is the most you want to have on the shelves, and you should not exceed that quantity. When you establish a minimum level (or par level), that is the point when you order more (commonly called your reorder point), but not till you reach that minimum quantity. Based on this information, look at the following examples below, including explanations about the appropriate number of vials that should be purchased.

Example: How many bottles of tetracycline 250 mg capsules should be ordered based on the following information?
Package size: 500 capsules/bottle
Minimum number of units: 120 capsules
Maximum number of units: 700 capsules
Current inventory level: 80 capsules

With this information in mind, only 1 bottle of tetracycline should be ordered as it has already dropped below the reorder point of 120 capsules, but two bottles would exceed the maximum number of units.

Example: How many bottles of metronidazole 500 mg tablets should be ordered based on the following information?
Package size: 50 tablets/bottle
Minimum number of units: 30 tablets
Maximum number of units: 120 tablets
Current inventory level: 12 tablets

With this information in mind, 2 bottles of metronidazole should be ordered as it has already dropped below the reorder point of 30 tablets, and two bottles will bring you as close to the maximum as possible without exceeding the maximum number of units.

Example: How many bottles of doxycycline 50 mg capsules should be ordered based on the following information?
Package size: 50 capsules/bottle
Minimum number of units: 30 capsules
Maximum number of units: 150 capsules
Current inventory level: 42 capsules

With this information in mind, no bottles of doxycycline should be ordered as it has not yet reached the reorder point.

Inventory requirements

While individual pharmacies may have various frequencies in which they verify their inventory levels, it is worth mentioning the requirements that the DEA sets for controlled substances. A pharmacy is required by the DEA to take an inventory of controlled substances every 2 years (biennially). This inventory must be done on any date that is within 2 years of the previous inventory date. The inventory record must be maintained at the registered location in a readily retrievable manner for at least 2 years for copying and inspection by the Drug Enforcement Administration. An inventory record of all Schedule II controlled substances must be kept separate from those of other controlled substances. Submission of a copy of any inventory record to the DEA is not required unless requested

When taking the inventory of Schedule II controlled substances, an actual physical count must be made. For the inventory of Schedule III, IV, and V controlled substances, an estimate count may be made. If the commercial container holds more than 1000 dosage units and has been opened, however, an actual physical count must be made.

State law may strengthen this requirement with annual actual physical counts of all controlled substances. It also may require such an inventory be submitted before reregistration by the board of pharmacy.

Quick review question 7

Which of the following is not an advantage of wholesaler purchasing?

   a. reduced turnaround time for orders
   b. lower inventory and associated costs
   c. higher purchase costs when compared to direct purchasing
   d. reduced commitment of time and staff when compared to direct purchasing

Quick review question 8

How are schedule III through V medications ordered for the pharmacy?

   a. DEA 222 triplicate form
   b. CSOS enabled software
   c. both a and b
   d. on a general warehouse order

Quick review question 9

How are schedule II medications ordered for the pharmacy?

a. DEA 222 triplicate form
b. CSOS enabled software
c. both a and b
d. on a general warehouse order

Quick review question 10

How many bottles of carisoprodol 350 mg tablets should be ordered based on the following information?
Package size: 24 tablets/bottle
Minimum number of units: 24 tablets
Maximum number of units: 100 tablets
Current inventory level: 48 tablets

a. 0 bottles
b. 1 bottle
c. 2 bottles
d. 3 bottles

Quick review question 11

How many bottles of levothyroxine 100 mcg tablets should be ordered based on the following information?
Package size: 100 tablets/bottle
Minimum number of units: 180 tablets
Maximum number of units: 720 tablets
Current inventory level: 140 tablets

a. 0 bottles
b. 5 bottles
c. 6 bottles
d. 580 tablets

Quick review question 12

How many bottles of risperidone 2 mg tablets should be ordered based on the following information?
Package size: 100 tablets/bottle
Minimum number of units: 120 tablets
Maximum number of units: 240 tablets
Current inventory level: 60 tablets

a. 0 bottles
b. 1 bottle

c. 2 bottles
d. 3 bottles

Quick review question 13

How many bottles of sildenafil citrate 100 mg tablets should be ordered based on the following information?
Package size: 30 tablets/bottle
Minimum number of units: 90 tablets
Maximum number of units: 210 tablets
Current inventory level: 69 tablets

a. 3 bottles
b. 4 bottles
c. 5 bottles
d. 6 bottles

Quick review question 14

How many bottles of glyburide 5 mg tablets should be ordered based on the following information?
Package size: 100 tablets/bottle
Minimum number of units: 200 tablets
Maximum number of units: 800 tablets
Current inventory level: 400 tablets

a. 0 bottles
b. 2 bottles
c. 4 bottles
d. 8 bottles

Quick review question 15

How often does the DEA require a pharmacy to count the inventory of their controlled substances?

a. semi-annually (every 6 months)
b. annually (every year)
c. biennially (every 2 years)
d. the DEA does not require controlled substances to be inventoried

Quick review question 16

Provided that any packages larger than 1000 dosage units have not been opened, which medication schedule(s) may a pharmacy estimate the count on when performing their biennial controlled substance inventory?

a. schedule III
b. schedule IV
c. schedule V
d. all of the above

Storage requirements

Now that we've discussed how to purchase and receive various pharmaceuticals, we need to start looking at how to properly store our inventory, as this is another area in which pharmacy technicians have great responsibility. There are three main concepts to discuss in this section:

- environmental considerations,
- security issues, and
- safety requirements.

Environmental considerations

Environmental considerations include proper temperature, ventilation, humidity, light and sanitation. Specific storage conditions are required to be printed in product literature, on drug packaging, and drug labels to ensure proper storage and product integrity. The conditions are defined by the following terms:

- Cold: any temperature not exceeding 8° C (45° F)
- Freezer: -25° to -10° C (-13° to 14° F)
- Refrigerator: 2° to 8° C (36° to 46° F)
- Cool: 8° to 15° C (46° to 59° F)
- Room temperature: the temperature prevailing in a working area
- Controlled room temperature: 15° to 30° C (59° to 86° F)
- Warm: 30° to 40° C (86° to 104° F)
- Excessive heat: any temperature above 40° C (104° F)

The temperatures that you will need to be most concerned with are freezer, refrigerator, and controlled room temperature. Pharmacies should maintain some sort of daily log for the refrigerators and freezers that medications are stored within.

Volatile or flammable substances, such as the alcohols that a pharmacy may use for compounding or other purposes, must be stored in an area with proper ventilation to prevent build up of fumes in case of accidental spill or damaged storage container.

Humidity can cause a tablet to become moist and powdery. While all medications should not be exposed to excessive levels of humidity, some medications, such as acyclovir, mycophenolate, and zidovudine, seem to be more sensitive to degradation from humidity.

There are more than 200 different medications which are light sensitive. The chemical composition of these medications can be altered by exposure to direct light. As an example, when nitroprusside is exposed to direct sun light, it will breakdown into cyanide. Some common light sensitive medications include acetazolamide, doxycycline, linezolid, and zolmitriptan. While many drugs need to be protected from light while in storage, their original package from the manufacturer should suffice. If you need to repackage any medications, always be sure to consult the manufacturers recommendations to determine if you need to place the medication in light resistant packaging or not.

Sanitation standards are usually set and enforced by the state that the pharmacy practices in.

Security issues

Security requirements that restrict access to medications to "authorized personnel only" are often the result of legal requirements, institutional policy, and established standards of practice. All drugs in an institutional setting must be maintained in restricted locations so that they are only accessible to professional staff who are authorized to receive, store, prepare, dispense, distribute, or administer such products. Whereas, in a community pharmacy, the public has ready access to various over the counter medications.

Prescription (legend) drugs require a prescription and are otherwise restricted to "authorized personnel only" such as pharmacists and pharmacy technicians in all pharmacy settings.

As should be expected, there are additional security measures with respect to controlled substances. Schedule III – V medications must either be stored in a secured vault or be distributed throughout the pharmacy stock. By dispersing your controlled substances throughout your inventory, you effectively prevent someone from being able to steal all your scheduled medications. Schedule II medications must also either be stored in a secured vault or be distributed throughout the pharmacy stock; although, some states specifically require Schedule II medications to be stored in a secured vault.

Safety requirements

The safety requirements include everything from the proper inventory rotation to avoid dispensing expired products, to material safety data sheets to provide the necessary information for safe clean up after accidental spills, to appropriate handling of oncology materials, and proper storage of chemicals and flammable items.

Proper rotation of inventory and periodic checking of expirations help to reduce the potential for dispensing expired medications. It also maximizes the utilization of inventory before medications become outdated. When looking at expirations on medication vials, it is important to note that if a medication only mentions the month and year but not the day, then you are to treat it as expiring at the end of the month. As an example, if a medication is marked as expiring on 02/2020, then you would treat it as expiring on February 29, 2020.

The Occupational Safety and Health Administration (OSHA) requires all work places, including pharmacies, to carry material safety data sheets (MSDS) for all hazardous substances that are stored on the premises. This includes oncology drugs and volatile chemicals, along with other hazardous chemicals. The MSDS provide handling, clean-up, and first-aid information.

Segregating inventory by drug categories helps to prevent potentially harmful errors. The Joint Commission (TJC), formerly known as the Joint Commission on Accreditation of Healthcare Organizations (JCAHO), requires that internal and external medications must be stored separately. This reduces the potential that someone will dispense or administer an external product for internal use. The Joint Commission also has requirements for separate storage of oncology drugs and volatile or flammable substances.

Hazardous drugs (i.e., oncology drugs) should have a separate space on the shelves and be labeled in such a way that it will alert staff of the hazardous potential of these medications. Oncology drugs are often cytotoxic themselves and must be handled with extreme care. They should be received in a sealed protective outer bag that restricts dissemination of the drug if the container leaks or is broken. When potential exists for exposure to hazardous drugs, all personnel involved must wear appropriate personal protective equipment while following a hazardous materials cleanup procedure. All exposed materials must be properly disposed of in hazardous waste containers.

Volatile or flammable substances (including tax free alcohol) require careful storage. They must have a cool location that is properly ventilated. Their storage area must be designed to reduce fire and explosion potential.

Quick review question 17

What is the minimum standard for how often refrigerator and freezer logs should be maintained if they are being used to store medications?

   a. continuously
   b. each shift
   c. daily
   d. weekly

Quick review question 18

According to the USP, what is an acceptable temperature range for a medication freezer?

   a. -25° to -10° C
   b. -13° to 14° C
   c. 2° to 8° C
   d. 36° to 46° C

Quick review question 19

According to the USP, what is an acceptable temperature range for a medication refrigerator?

a. -25° to -10° C
b. -13° to 14° C
c. 2° to 8° C
d. 36° to 46° C

Quick review question 20

Who is responsible for establishing the sanitation standards for pharmacy?

a. federal government
b. each state
c. DEA
d. OSHA

Quick review question 21

If an expiration on a manufacturer's label only mentions a month and year, which day of the month does the medication expire?

a. the first of the month
b. the last day of the month
c. the fifteenth of the month
d. Manufacturer's labels always list a specific day of the month.

Quick review question 22

If a medication were labeled as expiring on 10/2020, on what day does the medication expire?

a. October 1, 2020
b. October 31, 2020
c. November 1, 2020
d. 10/2020 is an illegal format for an expiration date on a medication, and should be listed as a class III drug recall.

Quick review question 23

Which government agency requires pharmacies to maintain material safety data sheets for all hazardous substances maintained on their premises?

a. FDA
b. DEA
c. TJC

d. OSHA

Product removal

One of the challenges in managing pharmacy inventory is properly dealing with product removal. Medications may need to be removed from stock because of drug recalls, medications that are either short-dated or expired, products damaged during shipping, products that are undesirable due to stock levels, and medications that were improperly stored. Pharmacies will need to utilize proper waste management and work with wholesalers, manufacturers, and reverse distributors to handle these challenges. We can break this into two broad categories: returnable and nonreturnable medications.

Returns

There are typically four reasons for returning medication inventory, and depending on the reason for making the return, there may be additional paperwork to fill out and process for these returns. Anyone on the pharmacy staff may prepare and process returns; typically, the only items that may need a pharmacist's signature are the controlled substances. The four categories of returns are excess stock, drug recalls, damaged inventory, and expired medication.

Excess stock

Pharmacies try to tightly control inventory to avoid excess stock, as an over abundance of a particular medication may be hard to distribute in a timely fashion and require storage space that may be better utilized by other inventory items. Excess stock may come from back orders that suddenly get fulfilled, accidental over-ordering, and changes in medication usage trends. The pharmacy may want to return this excess stock to the wholesaler. The wholesaler may charge the pharmacy a re-stocking fee.

Drug recalls

Drug recalls (as discussed in Chapter 2) may have varying degrees of potential harm, but when drug recalls occur, they are typically identified by the NDC and specific lot number of the affected products. These medications need to be removed from potential use in the pharmacy, whether stored in a secured area, or simply removed from stock and relabeled in a manner to assure that no one dispenses them. The pharmacy may also need to contact patients that received recalled medications. The drug recalls typically need to be returned to the manufacturer. The means of returning these medications may vary, as sometimes they may be returned to the manufacturer by the pharmacy, and other times, the pharmacy may send their returns via the wholesaler or through a reverse distributor.

Damaged inventory

Damaged inventory should be immediately removed from stock. If the inventory was damaged during transportation to the pharmacy, the wholesaler or manufacturer has a responsibility to provide for the return. If the damage is not discovered till after the delivery driver is gone it may be necessary to call and get an approval code for the return, and to make the delivery driver aware of the return for the next time they enter the pharmacy. If the medication is damaged once it is in the pharmacy, it is typically the pharmacy's responsibility; although, the pharmacy may still be able to get partial credit from wholesalers and/or manufacturers if the damage is the result of improper storage (i.e., a refrigerator that breaks down, and the temperature falls outside of the proper storage range for the drugs inside of it).

Expired medication

Expired and short-dated medications will need to be pulled from the pharmacy shelves to ensure that the patients do not receive these medications. Most pharmacies will have a policy to pull medications from the shelves at least three months in advance. If the medication is simply short-dated, the pharmacy may want to make sure others are aware of the date, and attempt to utilize that medication prior to it expiring. Short-dated OTC medications may be placed on clearance in the pharmacy. In some instances, the pharmacy may be able to get partial credit for short-dated medications provided the containers are intact and have not been opened.

Reverse distribution

In its simplest terms, reverse distribution is sending medications back to wholesalers and manufacturers. As there can be many challenges with managing all this, reverse distributors may broker these returns on behalf of the pharmacy. They will often manage all returns (including controlled substances) and waste management.

Nonreturnable medications and their disposal

Many things may not be returned to the wholesaler/manufacturer, including partial vials and bottles, reconstituted or compounded medications, drugs that have been repackaged by the pharmacy, and often, even expired drugs. Some medications, such as antibiotics, may still be disposed of in the trash, whereas other medications may need to be treated as hazardous waste and placed in an appropriate container to designate it as such. The pharmacy will need to contract an appropriately licensed waste management service to deal with these items.

Controlled substances have additional regulations with respect to their destruction. The DEA must be notified of these medications prior to their removal or destruction. Controlled substances must be counted and cosigned prior to their destruction, and the pharmacist is responsible for providing any return information required by the DEA. The DEA will issue a receipt for any schedule II medications that they destroy. This receipt must be kept for a minimum of five years with the pharmacy's schedule II inventory.

Quick review question 24

What are short-dated medications?

   a. medications that have not yet expired, but will expire soon
   b. medications that recently expired
   c. medications that cannot be kept long once reconstituted
   d. pediatric medications

Quick review question 25

Which of the following items are reverse distributors allowed to manage?

   a. controlled substances
   b. expired medications
   c. pharmacy waste

d. all of the above

Quick review question 26

Nonreturnable medications would usually include what items?

a. reconstituted medications
b. medications that have been repackaged by the pharmacy
c. partial vials
d. all of the above

Quick review question 27

Which items should be removed from the pharmacy stock?

a. expired medications
b. drug recalls
c. damaged inventory
d. all of the above

Quick review question 28

Which government agency must be notified if a pharmacy wishes to move or destroy controlled substances?

a. DEA
b. FDA
c. TJC
d. OSHA

Answers to quick review questions

1. B - Inventory is simply the entire stock on hand for sale at a given time.
2. C - A formulary is a list of medications available for use within a health care system.
3. A - Direct purchasing entails ordering medications directly from the original drug manufacturer.
4. D - Occupational Safety and Health Administration (OSHA) is a government agency within the United States Department of Labor responsible for maintaining safe and healthy work environments.
5. D - The National Drug Code is a unique product identifier

used for medications intended for human use. It is a unique 10-digit, 3-segment numeric identifier assigned to each medication. The first segment identifies the manufacturer. The second segment identifies a specific strength, dosage form, and formulation for a particular manufacturer. The third segment identifies package forms and sizes.

6. A - While some hazardous medications may cause photosensitivity, a medication that causes photosensitivity is not automatically hazardous. Hazardous drugs are drugs that are known to cause genotoxicity, which is the ability to cause a change or mutation in genetic material; carcinogenicity, which is the ability to cause cancer in animal models, humans or both; teratogenicity, which is the ability to cause defects on fetal development or fetal malformation; and lastly, hazardous drugs are known to have the potential to cause fertility impairment, which is a major concern for most clinicians.

7. C - Reduced turnaround time for orders, lower inventory and associated costs, and reduced commitment of time and staff are considered advantages to wholesaler purchasing. Higher purchase costs, when compared to direct purchasing, is a disadvantage of wholesaler purchasing.

8. D - Schedule III – V drugs may be ordered by a pharmacy or other appropriate dispensary on a general order from a wholesaler, and you should check the delivery in against the original order.

9. C - The purchase of schedule II controlled substances must be authorized by a pharmacist and executed on either a triplicate DEA 222 order form or an electronic 222 form through a controlled substances ordering system (CSOS).

10. A - No bottles of carisoprodol should be ordered, as it has not yet reached the reorder point.

11. B - Five bottles of levothyroxine should be ordered, as it has already dropped below the reorder point of 180 tablets, and five bottles will bring you as close to the maximum as possible without exceeding the maximum number of units.

12. B - One bottle of risperidone should be ordered, as it has already dropped below the reorder point of 120 tablets, and one bottle will bring you as close to the maximum as possible without exceeding the maximum number of units.

13. B - Four bottles of sildenafil citrate should be ordered, as it has already dropped below the reorder point of 90 tablets, and four bottles will bring you as close to the maximum as possible without exceeding the maximum number of units.
14. A - No bottles of glyburide should be ordered as it has not yet reached the reorder point.
15. C - A pharmacy is required by the DEA to take an inventory of controlled substances every 2 years (biennially).
16. D - For the inventory of Schedule III, IV, and V controlled substances, an estimate count may be made. However, if the commercial container holds more than 1000 dosage units and has been opened, an actual physical count must be made.
17. C - Pharmacies should maintain some sort of daily log for the refrigerators and freezers that medications are stored within.
18. A - A medication freezer should be maintained between -25° to -10° C (-13° to 14° F).
19. C - A medication refrigerator should be maintained between 2° to 8° C (36° to 46° F).
20. B - Sanitation standards are usually set and enforced by the state that the pharmacy practices in.
21. B - If the expiration date on a medication only mentions the month and year, but not the day, then you are to treat it as expiring at the end of the month.
22. B - If the expiration date on a medication only mentions the month and year, but not the day, then you are to treat it as expiring at the end of the month.
23. D - The Occupational Safety and Health Administration (OSHA) requires all work places, including pharmacies, to carry material safety data sheets (MSDS) for all hazardous substances that are stored on the premises. This includes oncology drugs and volatile chemicals, along with other hazardous chemicals.
24. A - A short-dated medication is one that will expire soon (in a short amount of time).
25. D - Reverse distributors will often manage all returns (including controlled substances), and waste management.
26. D - Many things may not be returned to the wholesaler/manufacturer, including partial vials and bottles,

reconstituted or compounded medications, drugs that have been repackaged by the pharmacy, and often, even expired drugs.

27. D - Medications may need to be removed from stock because of drug recalls, medications that are either short-dated or expired, products damaged during shipping, products that are undesirable due to stock levels, and medications that were improperly stored.

28. A - The DEA must be notified prior to the removal or destruction of controlled substances.

CHAPTER 8 Pharmacy Billing and Reimbursement

Key concepts

This chapter will cover the following knowledge areas to prepare you for the *Pharmacy Technician Certification Exam*:

- Reimbursement policies and plans (e.g., HMOs, PPO, CMS, private plans)
- Third-party resolution (e.g., prior authorization, rejected claims, plan limitations)
- Third-party reimbursement systems (e.g., PBM, medication assistance programs, coupons, and self-pay)
- Healthcare reimbursement systems (e.g., home health, long-term care, home infusion)
- Coordination of benefits

Terminology

To get started in this chapter, there are some terms that should be defined.

out-of-pocket - Out-of-pocket expenses are the costs that are considered the responsibility of the patient, including non-covered items, deductibles, and copays.

third-party payor - A third-party payor (also spelled payer) is an organization other than the patient (first party) or pharmacy/health care provider (second party) involved in the financing of personal health services including, but not limited to, prescriptions medication.

fraud - Fraud can be broadly defined as an act of deliberate deception performed to acquire an unlawful benefit. In a pharmacy, that may include billing for a medication or device the patient did not receive, or over billing for a medication or device that the patient did receive.

bank identification number (BIN) - On a health insurance card, a BIN is a six digit number used to identify a specific plan from a carrier making it easier for the PBM to process your prescription online. No actual bank is involved in this part of the process (the name is a hold over from early electronic banking terminology).

group number - A group number identifies your group, or business, from other groups, or businesses, who are insured by the same insurance company.

member number - A member number is a number in correlation to relationship to the family member that provides the insurance. Typically, the individual that is primarily insured has a member number of '00' or '01' and the next number is reserved for their spouse (even if the individual is not currently married). Therefore, the spouse would be either '01' or '02'. Then, any children that they provide insurance for continues consecutively from oldest to youngest.

Processor Control Number (PCN) - The Processor Control Number (PCN) is a secondary identifier for insurance that may be used in the routing of pharmacy transactions by the processor to aid in the receipt and adjudication of prescription claims. A PBM/processor/plan may choose to differentiate different plans/benefit packages with the use of unique PCNs. The PCN is an alphanumeric number defined by the PBM/processor, as this identifier is unique to their business needs. There is no official registry of PCNs.

private insurance - Private insurance plans are ones that consumers receive through their employer (or their family member's employer), or through individual purchases.

public insurance - Public insurance is insurance either provided by or subsidized by the government, such as Medicare and Medicaid.

managed care - Managed care (a term used to describe most health insurance policies) is used to describe a variety of techniques intended to reduce the cost of health benefits and improve the quality of care.

health maintenance organization (HMO) - A health maintenance organization (HMO) covers care provided by health professionals who have signed up with the HMO, and have agreed to treat patients according to the HMO's policies.

preferred provider organization (PPO) - A preferred provider organization (PPO) is an insurance plan in which participating health care professionals and hospitals treat patients for an agreed upon rate.

exclusive provider organization (EPO) - An exclusive provider organization (EPO) is an insurance plan in which only authorized health care professionals and hospitals treat patients for an agreed upon rate. This plan is usually more exclusive than a PPO.

Centers for Medicare & Medicaid Services (CMS) - The Centers for Medicare & Medicaid Services (CMS), previously known as the Health Care Financing Administration (HCFA), is a federal agency within the United States Department of Health and Human Services (DHHS) that administers the Medicare program and works in partnership with state governments to administer Medicaid, the State Children's Health Insurance Program (SCHIP), and health insurance portability standards.

Medicare - Federal health insurance program for patients 65 years old and above, some younger patients with disabilities, and some people with permanent kidney failure (end-stage renal disease).

Medicare Part A - Medicare Part A is the part of Medicare that pays for hospital care.

Medicare Part B - Medicare Part B is the part of Medicare that pays for doctor visits, certain injections, durable medical equipment, chemotherapy, and diabetes supplies (not insulin).

Medicare Part C - Medicare Part C, also called a Medicare Advantage Plan, is a type of Medicare health plan offered by a private company that contracts with Medicare to provide you with all your Part A and Part B benefits.

Medicare Part D - Medicare Part D is the part of Medicare that pays for prescription drug coverage. Medicare Part D covers outpatient prescription drugs exclusively through private plans or through Medicare Advantage plans that offer prescription drugs.

Medicaid - This is a joint federal and state program that helps with medical costs for some people with low incomes and limited resources. Medicaid programs vary from state to state. Some patients are covered by both Medicaid and Medicare.

National Council for Prescription Drug Programs (NCPDP) - The National Council for Prescription Drug Programs (NCPDP) is a nonprofit organization that create national standards for electronic health care transactions used in prescribing, dispensing, monitoring, managing and paying for medications and pharmacy services. The NCPDP also develop standardized business solutions and best practices that safeguard patients.

switch vendor - A switch vendor routes prescription information from the pharmacy management software to ensure that the information conforms to the NCPDP standards prior to routing it to the PBM.

prior authorization - If a medication is not normally covered by an insurance, is a particularly high dose, has significant risk potential, is being used to treat or ameliorate off-label disease(s)/condition(s), or is not usually recommended for a particular age or gender, the physician and/or the pharmacy may need to acquire prior authorization in order to get the insurance to cover the medication.

coordination of benefits - Determining which insurance should be considered primary, secondary, tertiary, etc. is sometimes referred to as coordination of benefits; although, that term is more commonly used if one of the insurance plans involve Medicare.

adjudication - Adjudication is a term used in the insurance industry to refer to the process of paying claims submitted or denying them after comparing claims to the benefit or coverage requirements.

remittance advice (RA) - Remittance advice (RA) is a document sent to the pharmacy by the insurance company providing the details of a paid claim. This is also referred to as an explanation of benefits (EOB).

fee-for-service reimbursement - Fee-for-service reimbursement is a payment method in which providers receive payment for each service rendered.

episode-of-care reimbursement - Episode-of-care reimbursement is a payment method in which providers receive one lump sum for all the services they provide related to a condition or disease.

Quick review question 1

What is the proper term used for when a pharmacy over charges for a medication that a patient receives?

a. fraud
b. human error
c. profit
d. insufficient data to properly answer this question

Quick review question 2

A Medicare Advantage Plan is the same as what?

a. Medicare Part A
b. Medicare Part B
c. Medicare Part C
d. Medicare Part D

Quick review question 3

A community pharmacy should process glucometer testing strips for a Medicare patient under which part of their insurance?

a. Part A
b. Part B
c. Part D
d. The patient will need to pay the cash price.

Quick review question 4

A community pharmacy should process insulin for a Medicare patient under which part of their insurance?

a. Part A
b. Part B
c. Part D
d. The patient will need to pay the cash price.

Quick review question 5

Why might a pharmacy or prescriber need to acquire prior authorization for a patient's medication to be accepted by their insurance?

a. The medication is not normally covered by the patient's insurance.
b. The medication has significant risk potential.
c. The medication prescribed is a particularly high dose.
d. all of the above

Quick review question 6

If a patient has multiple insurances, including a Medicare plan, what is the process referred to when determining which plan is considered primary and which plan is considered secondary?

a. coordination of benefits
b. PITA
c. primary and secondary insurance policy rule
d. none of the above

Quick review question 7

What is the name of the process when an insurance company either pays a claim or denies it?

a. billing
b. adjudication
c. remittance
d. none of the above

Quick review question 8

Which organization has created national standards for electronic health care transactions used in prescribing, dispensing, monitoring, managing, and paying for medications and pharmacy services.

a. DEA
b. FDA
c. ISMP
d. NCPDP

Health insurance

In an era of high medical costs (2012 saw an annual expenditure of $325.8 billion on prescription medications according to the IMS Institute for Healthcare Informatics), and the expectation of greater numbers of insured persons due to the individual mandate in the Patient Protection and Affordable Care Act (scheduled to take effect in January 2014), it is increasingly important to understand and properly process insurance claims. Pharmacies need to bill patients for their medications, medical supplies, and for services rendered (medication therapy management), and it is important to properly enter the necessary information and process claims correctly to avoid allegations of fraud.

Remuneration for pharmaceutical goods and counseling can be broken into three very broad groups: private insurance, public insurance, and cash. Private insurance plans are ones that consumers receive through their employer (or their family member's employer), or through individual purchases (which includes the online health exchanges mandated by the Patient Protection and Affordable Care Act). Public insurance is insurance either provided by or subsidized by the government, such as Medicare and Medicaid. Cash, while less common, is another option if the individual receiving the product either doesn't want to process it through their insurance, doesn't have insurance, or if the consumer still wants it after their insurance rejects it.

According to information published by the United States Census Bureau in 2011, roughly 55% of Americans obtain private insurance through an employer, while about 10% purchase it directly. About 31% of Americans were enrolled in a public health insurance program: 14.5% (45 million – although, that number has since risen to 48 million) had Medicare, 15.9% (49 million) had Medicaid, and 4.2% (13 million) had military health insurance (as there is some overlap with individuals having multiple plans, the percentages add up to more than 100%).

Private insurance health plans

Private health insurance may be purchased on a group basis (by a firm to cover its employees) or purchased by individual consumers. Most Americans with private health insurance receive it through an employer-sponsored program. According to information published by the United States Census Bureau in 2011, some 55% of Americans are covered through an employer sponsored program, while about 10% purchase health insurance directly. As additional portions of the Patient Protection and Affordable Care Act of 2010 come into effect, both of these numbers will likely rise due to the individual and employer mandates. In 2014, individuals not obtaining health insurance will receive a fine, and in 2015 employers with 50 or more full-time employees will also be fined if they do not offer health insurance to their full-time employees.

These health insurance plans may take a variety of forms, but most of them can be categorized as managed care. The term managed care is used to describe a variety of techniques intended to reduce the cost of health benefits and improve the quality of care. It is also used to describe organizations that use these techniques ("managed care organization"). Most managed care organizations are either health maintenance organizations (HMOs), preferred provider organizations (PPOs), or exclusive provider organizations (EPOs).

Public (government) health plans

Public health insurance programs provide the primary source of coverage for most seniors and for low-income children and families who meet certain eligibility requirements. The primary public programs are Medicare, a federal social insurance program for seniors (generally persons aged 65 and over) and certain disabled individuals; Medicaid, funded jointly by the federal government and states, but administered at the state level, which covers certain very low income children and their families; and State Children's Health Insurance Program (SCHIP), also a federal-state partnership that serves certain children and families who do not qualify for Medicaid, but who cannot afford private coverage. Other public programs include military health benefits provided through TRICARE and the Veterans Health Administration and benefits provided through the Indian Health Service. Some states have additional programs for low-income individuals.

Medicare

Medicare is a national social insurance program administered by the Centers for Medicare & Medicaid Services (CMS) that guarantees access to health insurance for Americans aged 65 and older, and younger people with disabilities, as well as people with end stage renal disease and individuals with ALS. In 2010, Medicare provided health insurance to 48 million Americans, 40 million people age 65 and older, and eight million younger people with disabilities.

Medicare offers all enrollees a defined benefit. Hospital care is covered under Part A, and outpatient medical services are covered under Part B. To cover the Part A and Part B benefits, Medicare offers a choice between an open-network single payer health care plan (traditional Medicare) and a network plan (Medicare Advantage, or Medicare Part C), where the federal government pays for private health coverage. According to 2012 statistics, 76% of Medicare enrollees have traditional Medicare, while 24% have enrolled in the Medicare Advantage plan. Medicare Part D covers outpatient prescription drugs exclusively through private plans or through Medicare Advantage plans that offer prescription drugs.

Medicaid

Medicaid is a government insurance program for people of all ages whose income and resources are insufficient to pay for health care. Medicaid is overseen by the Centers for Medicare and Medicaid Services. Medicaid is the largest source of funding for medical and health-related services for people with low income in the United States. It is a means-tested program that is jointly funded by the state and federal governments, and managed by the states, with each state currently having broad leeway to determine who is eligible for its implementation of the program. Medicaid recipients must be U.S. citizens or legal permanent residents, and may include low-income adults, their children, and people with certain disabilities. Poverty alone does not necessarily qualify someone for Medicaid.

The Patient Protection and Affordable Care Act will significantly expand both eligibility for and federal funding of Medicaid beginning on January 1, 2014. Under the law as written, all U.S. citizens and legal residents with income up to 133% of the poverty line, including adults without dependent children, would qualify for coverage.

However, the United States Supreme Court ruled in National Federation of Independent Business v. Sebelius that states do not have to agree to this expansion in order to continue to receive existing levels of Medicaid funding, and many states have chosen to continue with current funding levels and eligibility standards.

Other government sponsored programs

Besides Medicare and Medicaid, the government commonly provides assistance to a broad array of groups, including children and pregnant women (State Children's Health Insurance Program), military veterans (Veterans Health Administration), the families of military personnel (TRICARE), and native Americans (Indian Health Service).

SCHIP

The State Children's Health Insurance Program (SCHIP), more commonly known as the Children's Health Insurance Program (CHIP) is a joint state and federal program to provide health insurance to children in families who earn too much money to qualify for Medicaid, yet cannot afford to buy private insurance. This also includes prenatal care for women during their pregnancy, provided they meet the established financial criteria. SCHIP programs are run by the individual states according to requirements set by the federal Centers for Medicare and Medicaid Services (CMS), and may be structured as independent programs separate from Medicaid (separate child health programs), as expansions of their Medicaid programs (SCHIP Medicaid expansion programs), or combine these approaches (SCHIP combination programs).

TRICARE

TRICARE, formerly known as the Civilian Health and Medical Program of the Uniformed Services (CHAMPUS), is a health care program of the United States Department of Defense Military Health System. Tricare provides civilian health benefits for military personnel, military retirees, and their dependents, including some members of the Reserve Component. The Tricare program is managed by Tricare Management Activity (TMA) under the authority of the Assistant Secretary of Defense (Health Affairs). Tricare is the civilian care component of the Military Health System (although historically it also included health care delivered in the military medical treatment facilities).

CHAMPVA

Civilian Health and Medical Program of the Department of Veterans Affairs (CHAMPVA) is a health benefits program managed by the Department of Veterans Affairs (VA) for eligible beneficiaries. Eligible beneficiaries include the spouse or widow(er), and to the children of a veteran who is rated permanently and totally disabled due to a service-connected disability, or was rated permanently and totally disabled due to a service-connected condition at the time of death, or died of a service-connected disability, or died on active duty and the dependents are not otherwise eligible for TRICARE benefits.

Veterans Health Administration

The Veterans Health Administration (VHA) is the component of the United States Department of Veterans Affairs (VA) that implements the medical assistance program of the VA through the administration and operation of numerous VA outpatient clinics, hospitals, medical centers and long-term healthcare facilities (i.e., nursing homes). By Federal law, eligibility for benefits is determined by a system of eight Priority Groups. Retirees from military service, veterans with service-connected injuries or conditions rated by VA, and Purple Heart recipients are within the higher priority groups.

Indian Health Service

The Indian Health Service (IHS) is an operating division (OPDIV) within the U.S. Department of Health and Human Services (HHS). IHS is responsible for providing medical and public health services to members of federally recognized Tribes and Alaska Natives. IHS is the principal federal health care provider and health advocate for Indian people, and its goal is to raise their health status to the highest possible level.

Workers' compensation

Workers' compensation is a form of insurance providing wage replacement and medical benefits to employees injured in the course of employment. Workers' compensation benefits vary according to state law.

Quick review question 9

What do pharmacies commonly bill insurances for?

a. medication
b. medical supply
c. services rendered (i.e., MTM)
d. all of the above

Quick review question 10

Which of the following is not considered managed care?

a. cash
b. HMO
c. PPO
d. EPO

Quick review question 11

Which organization oversees both Medicare and Medicaid?

a. FDA
b. DEA
c. CMS
d. OSHA

Quick review question 12

Which part of Medicare covers outpatient prescriptions?

a. Part A
b. Part B
c. Part C
d. Part D

Quick review question 13

Where do the funds for Medicaid come from?

a. state government
b. federal government
c. state and federal governments
d. none of the above

Quick review question 14

Besides patients enrolled in Medicare and Medicaid, what other groups commonly receive medical insurance from the government?

a. children and pregnant women
b. military veterans and their families
c. native Americans
d. all of the above

Quick review question 15

Which insurance is intended for children in families who earn too much money to qualify for Medicaid, yet cannot afford to buy private insurance?

a. Indian Health Services
b. SCHIP
c. CHAMPVA
d. workers' compensation

Quick review question 16

Which of the following is not a government funded health insurance?

a. Social Security
b. Medicare
c. Medicaid
d. SCHIP

Quick review question 17

The individual mandate in the Patient Protection and Affordable Care Act is expected to increase the number of people that purchase what kind of insurance?

a. no insurance
b. private insurance
c. public insurance
d. home owner's insurance

Quick review question 18

The employer mandate in the Patient Protection and Affordable Care Act applies to which businesses?

a. all businesses
b. businesses with more than 15 employees
c. businesses with 50 or more employees
d. businesses with 50 or more full-time employees

Pharmacy billing cycle

The pharmacy billing pathway in most community pharmacies can be broken into the following steps:

1. Receiving the prescription
2. Gathering patient data
3. Data entry
4. Pharmacy claim transmittal
5. Third-party payor adjudication
6. Point-of-sale
7. Payment processing

While parts of this process were already covered in *Chapter 6: Medication Order Entry and Fill Process*, we will be focusing on the correlation between these processes and pharmacy billing/reimbursement.

Receiving the prescription

When a community pharmacy receives a prescription, it is a requirement to note the source of the prescription if it is for a Medicare or Medicaid patient. As many individual insurance companies also require this information, it has become a common practice for community pharmacies to track where all prescriptions come from. These prescriptions are tracked using prescription origin codes (POC), which are commonly entered into the pharmacy management software. The prescription origin codes are as follows:

0 = Unknown is used when the manner in which the original prescription was received is not known, which may be the case in a transferred prescription.
1 = Written prescription via paper, which includes computer printed prescriptions that a physician signs as well as tradition prescription forms
2 = Telephone prescription obtained via oral instruction or interactive voice response
3 = E-prescriptions securely transferred from a computer to the pharmacy
4 = Facsimile prescription obtained via fax transmission, including an

e-Fax where a scanned image is sent to the pharmacy, and either printed or displayed on a monitor/screen

Gathering patient data

When a patient first arrives at a community pharmacy, the pharmacy team will need to gather/verify various important pieces of patient data including:

- gather drug and disease information,
- ensure that the pharmacy has the correct name, address, date of birth, contact information, and any other pertinent data,
- document/update allergy information,
- verify/update medication insurance information, which includes coverage type (primary, secondary, etc.), insurance name and bank identification number (BIN), group number, and member number.

Data entry

Pharmacy staff will need to enter information into the pharmacy software management system in order to process the prescription. Common information to require includes:

- *Prescriber information* -This typically includes the prescriber's name, address of practice, contact information, medical license number, DEA number, and National Provider Identifier (NPI).
- *Third-party payor* - This includes coverage type (primary, secondary, etc.), insurance name and bank identification number (BIN), group number, and member number. If a patient has multiple insurance plans, be sure to enter them correctly as primary, secondary, etc.
- *Patient information* - The patient information should at least include name, date of birth, address, contact information, allergies, and payment type (cash vs. insurance). Often, pharmacies will request information on concurrent use of other medications and dietary supplements, preferences

with respect to safety lids, and verification that the patient has received notification of the pharmacy's privacy policy.

- *Prescription information* - While many items on a prescription are important, the system should record as a minimum the date the prescription was written, superscription, inscription, subscription, signatura, refills, prescription origin code, and it should generate a unique prescription number that should appear on the prescription label as well.
- *DAW codes* - Dispense as written (DAW) codes need to be entered into the computer as well. Most prescriptions allow for generic substitution, and patients are glad to receive the more affordable version; therefore, the default DAW code is typically set to '0'. If a physician requires a specific medication to be dispensed, they will typically note this on the prescription. This is considered a DAW code of '1'. Sometimes a patient may request that they receive a brand name product even if a prescriber allowed for generic substitution. This would be classified as a DAW code of '2'. Other DAW codes are less frequently used. The following is a succinct list of the other DAW codes; 3 = substitution allowed - pharmacist selected product dispensed, 4 = substitution allowed - generic drug not in stock, 5 = substitution allowed - brand drug dispensed as generic, 6 = override, 7 = substitution not allowed - brand drug mandated by law, 8 = substitution allowed - generic drug not available in marketplace, and 9 = Other.
- *Drug information* - At a minimum, drug information should include the drug name, the medication's National Drug Code (NDC), the manufacturer, and an ability to check for interactions and contraindications. Often this drug information will include information on auxiliary labels, specific lot numbers and expiration dates, stock availability, pricing, and medication guides.

Pharmacy claim transmittal

At this point, the pharmacy is ready to transmit the prescription. This process is diagrammed below. When the prescription transmits, it goes through the switch vendor, and is either accepted (approved) or sent on to the PBM. If declined, the pharmacy, the prescriber, and/or the patient will need to contact either the PBM, or the third-party payor to attempt to obtain approval. If a patient has multiple insurance plans, most pharmacy management software systems are capable of performing split-billing. Determining which insurance should be considered primary, secondary, tertiary, etc. is sometimes referred to as coordination of benefits; although, that term is more commonly used if one of the insurance plans involve Medicare.

Third Party Claim Pathway for Community Pharmacy

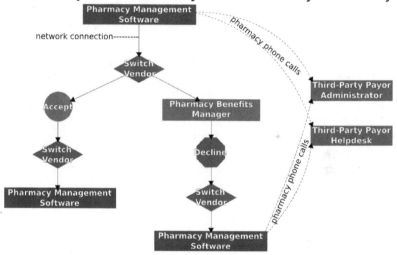

Common reasons for rejections include:

- a noncovered medication, or a medication requiring prior authorization,
- incorrect days' supply of medication, refill too soon, or invalid quantity of medication being dispensed,
- the patient's insurance on file is not currently active, or has been incorrectly entered, and
- the prescriber's information is either incomplete or entered incorrectly.

To assist in the resolution of declined prescriptions, the NCPDP has established a standardized set of reject codes. Some common examples include:

01 = Missing/Invalid BIN
09 = Missing/Invalid Birth Date
11 = Missing/Invalid Relationship Code
19 = Missing/Invalid Days Supply
25 = Missing/Invalid Prescriber ID
66 = Patient Age Exceeds Maximum Age
70 = Product/Service Not Covered
75 = Prior Authorization Required
79 = Refill Too Soon

Third-party payor adjudication

Once the prescription is accepted, the claim is adjudicated by the payor. The payor compares the charges with the terms of the patient's benefit plan, and determines what the patient owes, as well as what the insurance plan is financially responsible for. This information is then returned to the pharmacy electronically. Because this transaction process occurs in a matter of seconds, it is sometimes referred to as real-time claim adjudication (RTCA).

Point-of-sale

Once the medication has been filled and checked, it is ready for the patient to pick it up, which usually involves any additional payments (copays, deductibles, or if a particular medication is not covered, then the usual and customary price), the offering of medication counseling, and the proper recording and filing of dispensed prescriptions. Common payment options include coupons from the manufacturers, cash, checks, credit cards, and debit cards. Often, the pharmacy management system can handle this portion of the prescription filling process as well.

Payment processing

The insurance companies will send out payments (electronically or by paper check) to the pharmacies every thirty to sixty days for all prescriptions processed within a particular time frame. This payment is typically accompanied by a remittance advice (RA), also known as an explanation of benefits, providing details about the paid claims.

Quick review question 19

If a pharmacy receives a prescription on a traditional paper prescription form, which POC should be entered into the pharmacy management software?

a. 1
b. 2
c. 3
d. 4

Quick review question 20

If a physician leaves a prescription order on the pharmacy's IVR, which POC applies to that?

a. 1
b. 2
c. 3
d. 4

Quick review question 21

When entering a physician into your pharmacy management software, which number(s) will you likely need to process the prescription through the patient's insurance?

a. medical license number
b. DEA number
c. NPI
d. all of the above

Quick review question 22

When transmitting a claim from the pharmacy management software to the insurance, what is the first entity that reviews the claim?

a. third-party payor
b. PBM
c. switch vendor

d. BIN

One of the most common reject codes to receive in community pharmacy is '79'. What does this reject code mean?

a. refill too soon
b. missing/invalid prescriber ID
c. missing/invalid days supply
d. switch vendor is currently down

Which organization established a standardized set of reject codes?

a. FDA
b. switch vendors
c. PBMs
d. NCPDP

What is another term for remittance advice?

a. bill
b. explanation of benefits
c. payment received
d. counseling

When an insurance company sends a remittance advice, what does it provide?

a. It warns that if corrective measures are not taken your pharmacy will be sent to collections.
b. It provides details about the claims paid.
c. It provides a reminder of money the pharmacy owes to the insurance.
d. It details which patients would be prime candidates for MTM.

Other healthcare reimbursement methodologies

So far, this chapter has focused on third-party payor reimbursement in community pharmacy settings. Healthcare reimbursement often works differently in other practice settings (home health, long-term care, home infusion, etc.). The two major types of payment in these practice settings include fee-for-service reimbursement or episode-of-care reimbursement.

Fee-for-service reimbursement is a payment method in which providers receive payment for each service rendered. A fee is a set amount or a set price. Fee-for-service means a specific payment is made for each specific service rendered. In the fee-for-service method, the provider of the healthcare service charges a fee for each type of service, and the health insurance company pays each fee for a covered service. These fees or prices are known as charges in healthcare. Typically, the pharmacy or healthcare organization bills for each service provided on claims that have specific values associated with them. The claim is submitted to the third party payor, and the insurance company will send out payments covering windows of time and include remittance advices.

Episode-of-care reimbursement is a payment method in which providers receive one lump sum for all the services they provide related to a condition or disease. In the episode-of-care payment method, the unit of payment is the episode, not each individual health service. Therefore, the episode-of-care payment method eliminates individual fees or charges. The episode-of-care payment method controls costs on a systematic scale. If a specific amount is contracted to the pharmacy, this may also be referred to as a capitation fee.

Quick review question 27

Which of the following terms best describes a payment method in which providers receive payment for each service rendered?

a. fee-for-service
b. episode-of-care
c. fraud

d. highway robbery

Quick review question 28

Which of the following terms best describes a payment method in which providers receive one lump sum for all the services they provide related to a condition or disease?

a. fee-for-service
b. episode-of-care
c. fraud
d. highway robbery

Answers to quick review questions

1. A - Fraud can be broadly defined as an act of deliberate deception performed to acquire an unlawful benefit. In a pharmacy, that may include billing for a medication or device the patient did not receive, or over billing for a medication or device that the patient did receive.
2. C - Medicare Part C, also called a Medicare Advantage Plan, is a type of Medicare health plan offered by a private company that contracts with Medicare to provide you with all your Part A and Part B benefits.
3. B - Medicare Part B is the part of Medicare that pays for doctor visits, certain injections, durable medical equipment, chemotherapy, and diabetes supplies (not insulin).
4. C - Medicare Part D is the part of Medicare that pays for prescription drug coverage. Medicare Part D covers outpatient prescription drugs exclusively through private plans, or through Medicare Advantage plans that offer prescription drugs.
5. D - If a medication is not normally covered by an insurance, is a particularly high dose, has significant risk potential, is being used to treat or ameliorate off-label disease(s)/condition(s), or is not usually recommended for a particular age or gender, the physician and/or the pharmacy may need to acquire prior authorization in order to get the insurance to cover the medication.
6. A - Determining which insurance should be considered

primary, secondary, tertiary, etc. is sometimes referred to as coordination of benefits; although, that term is more commonly used if one of the insurance plans involve Medicare.

7. B - Adjudication is a term used in the insurance industry to refer to the process of paying claims submitted, or denying them after comparing claims to the benefit or coverage requirements.

8. D - The National Council for Prescription Drug Programs (NCPDP) is a nonprofit organization that create national standards for electronic health care transactions used in prescribing, dispensing, monitoring, managing, and paying for medications and pharmacy services. The NCPDP also develop standardized business solutions and best practices that safeguard patients.

9. D - Pharmacies will need to bill for medications, medical supplies, and for services rendered (medication therapy management).

10. A - Most managed care organizations are either health maintenance organizations (HMOs), preferred provider organizations (PPOs), or exclusive provider organizations (EPOs).

11. C - The acronym CMS stands for the Centers for Medicare & Medicaid Services.

12. D - Medicare Part D is the part of Medicare that pays for prescription drug coverage. Medicare Part D covers outpatient prescription drugs exclusively through private plans, or through Medicare Advantage plans that offer prescription drugs.

13. C - Medicaid is jointly funded by the state and federal governments.

14. D - All those groups often receive government funded health insurance through SCHIP, TRICARE, CHAMPVA, Veterans Health Administrations, and IHS.

15. B - SCHIP is a joint state and federal program to provide health insurance to children in families who earn too much money to qualify for Medicaid, yet cannot afford to buy private insurance.

16. A - Social Security is the Old-Age, Survivors, and Disability Insurance (OASDI) federal program, but does not include

health insurance. Medicare, Medicaid, and SCHIP are government funded health insurances.

17. B - The Patient Protection and Affordable Care Act is expected to increase the number of people who purchase private insurance, and the number of people who receive public insurance.

18. D - The employer mandate will apply to businesses with 50 or more full-time employees.

19. A - A POC of 1 applies to written prescription via paper, which includes computer printed prescriptions that a physician signs as well as tradition prescription forms.

20. B - A POC of 2 applies to telephone prescription obtained via oral instruction or interactive voice response.

21. D - The required prescriber information typically includes the prescriber's name, address of practice, contact information, medical license number, DEA number, and National Provider Identifier (NPI).

22. C - When the prescription transmits, it goes through the switch vendor, and is either accepted (approved) or sent on to the PBM.

23. A - A reject code of '79' means that the PBM thinks that the refill is being processed too soon.

24. D - To assist in the resolution of declined prescriptions the NCPDP has established a standardized set of reject codes.

25. B - A remittance advice (RA), also known as an explanation of benefits, provides details about the paid claims.

26. B - A remittance advice (RA), also known as an explanation of benefits, provides details about the paid claims.

27. A - Fee-for-service reimbursement is a payment method in which providers receive payment for each service rendered.

28. B - Episode-of-care reimbursement is a payment method in which providers receive one lump sum for all the services they provide related to a condition or disease.

CHAPTER 9 Pharmacy Information System Usage and Application

Key concepts

This chapter will cover the following knowledge areas to prepare you for the *Pharmacy Technician Certification Exam*:

- Pharmacy-related computer applications for documenting the dispensing of prescriptions or medication orders (e.g., maintaining the electronic medical record, patient adherence, risk factors, alcohol drug use, drug allergies, side effects)
- Databases, pharmacy computer applications, and documentation management (e.g., user access, drug database, interface, inventory report, usage reports, override reports, diversion reports)

Terminology

To get started in this chapter, there are some terms that should be defined.

database - A database is an organized collection of data.

database management system (DBMS) - A database management system (DBMS) is a software system designed to allow the definition, creation, querying, update, and administration of databases.

query - In computing, a query is a precise request for information retrieval with database and information systems.

health information exchange (HIE) - Health information exchange

(HIE) is the mobilization of healthcare information electronically across organizations within a region, community or hospital system.

pharmacy informatics - Pharmacy informatics (a.k.a., pharmacoinformatics) is the use and integration of data, information, knowledge, technology, and automation in the medication use process. The goal of this integration is to improve health outcomes.

reconciliation - In pharmacy, the term reconciliation can have two different definitions. Reconciliation can involve correcting a pharmacy charge to correspond with the proper fees for the item dispensed. Reconciliation can mean evaluating the patient's response to a therapy and making the necessary changes to better ensure that the patient's medical treatment better coincides with their needs.

cart-fill - Cart-fill (sometimes listed as cartfill) is the process of providing patients with a short term supply of all the medications they need (typically 24 hours worth). This process is commonly practiced in institutional settings.

computer cracker - A computer cracker is someone who makes unauthorized use of a computer, especially to tamper with data or programs. The media commonly refers to these individuals as hackers.

Quick review question 1

What does pharmacy informatics make use of?

  a. data
  b. technology
  c. automation
  d. all of the above

Quick review question 2

Which of the following terms is often used by the media to describe a computer cracker?

  a. hacker
  b. black hat specialist
  c. a Caucasian nerd
  d. byte sized sets of data that can fit on a thin crisp wafer

Databases

Pharmacy information systems are providing the backbone for today's plethora of pharmacy automation and technology, as they need to be able to access vast quantities of information to perform their necessary functions. These databases usually require someone with a strong understanding of how technology actually works to be properly implemented as they will need to determine what kind of database management system to use (i.e., MySQL/MariaDB, PostgreSQL, MS SQL, Oracle Database, etc.) and how they want information to be added (entering it directly into the database, other software interfaces that connect to the database being allowed to write to the database, replicating existing databases and maintaining updates against the other databases, or by other means).

This field is commonly referred to as pharmacoinformatics.

How a database works

The common database interactions can be broken into four main groups:

1. Data definition - Defining new data structures for a database, removing data structures from the database, modifying the structure of existing data.
2. Update - Inserting, modifying, and deleting data.
3. Retrieval - Obtaining information either for end-user queries and reports, or for processing by applications.
4. Administration - Registering and monitoring users, enforcing data security, monitoring performance, maintaining data integrity, dealing with concurrency control, and recovering information if the system fails.

Advantages

As these databases become more robust, they are able to readily organize large quantities of data that can be easily queried. This information is quickly retrievable, and databases do not tire of repetitive search strings, achieving a higher degree of accuracy and speed than what could be achieved without them.

Challenges

A number of challenges exist with creating and maintaining databases. A common challenge is getting the information from one database to communicate with a different database (especially if they are using different database management systems). A database is not generally portable across different database management systems, but different database management systems can interoperate by using standards such as SQL and ODBC or JDBC to allow a single application to work with more than one database. Health information exchanges have been developed to help with this process.

Security is another significant challenge, as many applications may need to be able to query information from these databases while still keeping unauthorized individuals from accessing the information. Database access control deals with controlling who (a person or a certain computer program) is allowed to access what information in the database. The information may comprise specific database objects (e.g., record types, specific records, data structures), certain computations over certain objects (e.g., query types, or specific queries), or utilizing specific access paths to the former (e.g., using specific indexes or other data structures to access information). Database access controls are set by special authorized (by the database owner) personnel that uses dedicated protected security database management system interfaces. This may be managed directly on an individual basis, or by the assignment of individuals and privileges to groups, or (in the most elaborate models) through the assignment of individuals and groups to roles which are then granted entitlements. Data security prevents unauthorized users from viewing or updating the database. Using passwords, users are allowed access to the entire database or subsets of it called "subschemas". For example, an employee database can contain all the data about an individual employee, but one group of users may be authorized to view only payroll data, while others are only allowed access to work with history and medical data. The database management system provides a way to interactively enter and update the database, as well as interrogate it.

This capability allows for database management.

Quick review question 3

What are MySQL, MS SQL, and Oracle Database examples of?

a. databases
b. database management systems
c. electronic health records
d. telemedicine

Quick review question 4

Which of the following is an advantage of a database?

a. it can organize large data sets
b. it can readily retrieve data
c. it can perform with a high degree of accuracy
d. all of the above

Quick review question 5

Which of the following statement about databases is false?

a. Different database management systems can interoperate by using standards such as SQL and ODBC or JDBC.
b. Database access control deals with controlling who is allowed to access what information in the database.
c. Databases can readily exchange information directly.
d. Databases do not tire of repetitive search strings.

Quick review question 6

Other than the database administrator(s), most people will only be able to access portions of the database. What is the technical term for these data subsets?

a. formatting
b. subschemas
c. encryption
d. data subsets

Medication use process

Pharmacy information systems have become so common and well integrated that they can directly impact every step of the medication use process. The following is a brief flowchart explaining the medication use process.

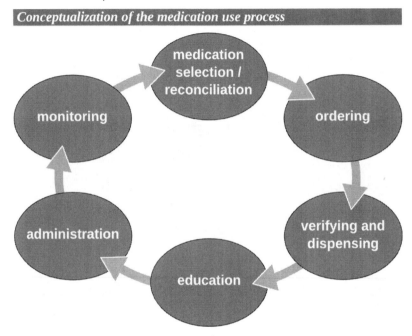

Conceptualization of the medication use process

Medication selection

Prescribers have become accustomed to consulting databases in their electronic health records when selecting medications in order to assure that the medications are covered, or to seek prior authorization if they deem a non covered medication to be necessary.

Ordering

Physicians, whether seeing a patient at their practice or rounding at an institution, commonly utilize e-prescribing and/or computerized prescriber order entry (CPOE) to order medications from the pharmacy. Physicians may also place orders using the pharmacy's interactive voice response (IVR) system.

Verifying and dispensing

In the case of CPOE, pharmacies only need to verify the medication and the appropriateness of the prescription (the pharmacy management software will help to screen for various interactions). E-prescribing commonly connects to modern pharmacy management systems; although, the pharmacy may still need to enter the prescription information into the system in a different manner to process it.

In both community pharmacy and institutional pharmacy, it has become common to scan barcodes on medications prior to them being dispensed to ensure accuracy.

In community pharmacy, the pharmacy may use an electronic signature capture to verify who picked the medication up, and to verify whether or not the patient desires counseling. Also, in some communities, pharmacy kiosks have begun to appear providing a convenient means of getting a common medication filled, or to pick-up a medication refill.

In many institutions, automated dispensing cabinets, robotic delivery systems, and/or pneumatic tube systems may be utilized to dispense the medications to the nursing units.

Education

The clinical decision support (CDS) tools allow for easy access to patient education points and guides including applicable warnings, contraindications, and side effects. On some medications where medication guides are mandatory, the pharmacy management systems will often prompt the pharmacy to provide the necessary information.

Telemedicine and telepharmacy have been implemented in many places, providing greater access for patients located either off-site or in rural areas to health care providers. This increased availability allows for improved patient education opportunities as well.

Administration

In outpatient settings, there has been a number of niche technology products intended to help patients take their medications in a timely fashion, ranging from simple alarms all the way to automated pill dispensing boxes that the pharmacy may program for the patient.

On the side of institutional medications, the integrated hospital software will remind nurses when to give patients their medications. Patient bedside barcode scanners will assure that the right medication is being given to the right patient at the right time, and often, even provide the right documentation by automatically charting the medication as being given.

Monitoring

Pharmacy technology exists in both inpatient and outpatient settings to monitor patient adherence by recording when medications have been given (or missed), and to monitor biological data such as blood sugar, blood pressure, etc. As the technologies to record this information continue to evolve, expect to see more of this information recorded on the patients' electronic health records.

Reconciliation

By accumulating and disseminating this information, it allows the entire health care team, including pharmacy, to make better decisions about what changes may need to be done to improve the patient's quality of life.

Quick review question 7

Which step within the medication use process is not impacted by pharmacy informatics?

a. prescription ordering
b. patient education
c. patient monitoring
d. all of the steps are impacted by pharmacy informatics

Quick review question 8

Prescribers will commonly order outpatient prescription using what type of software?

a. e-prescription
b. CPOE
c. DBMS
d. smoke signals

Quick review question 9

Prescribers will commonly order inpatient prescription using what type of software?

a. e-prescription
b. CPOE
c. DBMS
d. Rube Goldberg device

Quick review question 10

In institutional settings, bedside barcode scanners could be used to help assure all of the following except:

a. right drug
b. right attitude
c. right patient
d. right time

Quick review question 11

Thanks to advances in medical device technologies, health care can remotely monitor what?

a. medication adherence
b. blood sugar
c. blood pressure
d. all of the above

Quick review question 12

Many pharmacies today receive oral prescription messages through recordings made on their telephone system. What is the common acronym for these phone systems.

  a. DBMS
  b. ACD
  c. IVR
  d. PhBX

Specific pharmacy automations and technologies

Many technologies have already been named in this chapter, but this section will explain a little bit more about each of these technologies. As you read this section, it will be easy to see how technology can improve clinical decisions and outcomes, speed, accuracy, space utilization, and inventory management.

EHR

An electronic health record (EHR) is a collection of electronic health information about individual patients. It is a record in digital format that is theoretically capable of being shared across different health care settings. This sharing often occurs by way of network-connected, enterprise-wide information systems and other information networks or exchanges. These exchanges are sometimes referred to as health information exchanges (HIEs). EHRs may contain a range of data, including demographics, medical history, medication and allergies, immunization status, laboratory test results, radiology images, vital signs, personal statistics (like age and weight), and billing information.

As EHRs are still rapidly evolving, not all information has been transferred into them. Working backwards, there will be challenges gathering and entering all pertinent information from the past if it is not readily retrievable, but this is a short term problem, as all data moving forward should be entered into the necessary databases.

E-prescribing

E-prescribing is the computer-based electronic generation, transmission, and filling of a medical prescription, taking the place of paper and faxed prescriptions. E-prescribing allows a physician, nurse practitioner, or physician assistant to electronically transmit a new prescription or renewal authorization to a community or mail-order pharmacy.

E-prescribing provides the following benefits:

- Improving patient safety and quality of care - Illegibility from handwritten prescriptions is eliminated, decreasing the risk of medication errors. Also, oral miscommunication regarding prescriptions can be reduced, as e-prescribing should decrease the need for phone calls between prescribers and dispensers.
- Reducing time spent on phone calls and call backs to pharmacies - Many written prescriptions can be ambiguous and require pharmacies to call the prescriber, leave a message, and wait for a call back. This translates into less time available to the pharmacist for other important functions. E-prescribing can significantly improve this situation by reducing the volume of pharmacy call backs related to illegibility, mistaken prescription choices, formulary and pharmacy benefits.
- Automating the prescription renewal request and authorization process - With e-prescribing, renewal authorization can be an automated process that provides efficiencies for both the prescriber and pharmacist. Pharmacy staff can generate a renewal request (authorization request) that is delivered through the electronic network to the prescriber's system. The prescriber can then review the request, and act accordingly by approving or denying the request through updating the system. With limited resource utilization, and just a few clicks on behalf of the prescriber, they can complete a medication renewal and maintain continuous patient documentation.
- Increasing patient convenience and medication compliance - It is estimated that 20% of paper-based prescription orders

go unfilled by the patient, partly due to the hassle of dropping off a paper prescription and waiting for it to be filled. By eliminating or reducing this waiting period, e-prescribing may help reduce the number of unfilled prescriptions and increase medication compliance. Allowing the renewal of medications through this electronic system also helps improve the efficiency of this process, reducing obstacles that may result in less patient compliance. Availability of information on when patient's prescriptions are filled can also help clinicians assess patient compliance.

- Improving formulary adherence permits lower cost drug substitutions - By checking with the patient's health plan or insurance coverage at the point of care, generic substitutions or lower cost therapeutic alternatives can be encouraged to help reduce patient costs. Lower costs may also help improve patient compliance.
- Allowing greater prescriber mobility - Improved prescriber convenience can be achieved when using mobile devices that work on a wireless network, to write and renew prescriptions. Such mobile devices may include laptops, tablets, and smart phones. This freedom of mobility allows prescribers to write/renew prescriptions anywhere, even when not in the office.
- Improving drug surveillance/recall ability - E-prescribing systems enable embedded, automated analytic tools to produce queries and reports, which would be close to impossible with a paper-based system. Common examples of such reporting would be: finding all patients with a particular prescription during a drug recall, or the frequency and types of medication provided by certain health care providers.

CPOE

Computerized prescriber order entry, also called computerized physician order entry, is a process of electronic entry of practitioner instructions for the treatment of patients under their care. CPOE systems are used for processing orders in institutional settings.

The following is a list of features associated with CPOE:

- Customized-standardized ordering sets - Physician orders are standardized across the organization, yet may be individualized for each prescriber or specialty by using order sets. Orders are communicated to all departments and involved caregivers, improving response time and avoiding scheduling problems and conflict with existing orders.
- Patient-centered decision support - The ordering process includes a display of the patient's medical history and current results, and evidence based clinical guidelines to support treatment decisions.
- Patient safety features - The CPOE system allows real-time patient identification, drug dose recommendations, adverse drug reaction reviews, and checks on allergies and test or treatment conflicts. Physicians, nurses, and pharmacists can review orders immediately for confirmation.
- Regulatory compliance and security - Access is secure, and a permanent record is created with electronic signature.
- Portability - The system accepts and manages orders for all departments at the point-of-care, from any location in the health system (physician's office, hospital or home) through a variety of devices, including PCs, tablets, smart phones, and other devices.
- Health system reporting - The system delivers statistical reports online so that managers can analyze patient census and make changes in staffing, replace inventory, and audit utilization and productivity throughout the organization. Data is collected for training, planning, and root cause analysis for patient safety events.
- Billing accuracy - Documentation is improved by linking diagnoses to orders at the time of order entry to support appropriate charges.

IVR

Interactive voice response (IVR) is a technology that allows a computer to interact with people through the use of voice and dial tones. IVR applications can be used to control almost any function where the interface can be broken down into a series of simple interactions. In pharmacy, these systems are commonly used to direct phone calls, and order refills; physicians will frequently use them to phone in new prescriptions and/or clarifications.

Pharmacy management software systems

Pharmacy management software systems are the all-in-one tool that many community pharmacies depend on. Pharmacy management systems are usually capable of performing the following tasks:

- prescription processing
- e-prescribing
- prescription pick-up with signature capture
- refill queue
- work flow manager
- prescription scanning and patient card scanning
- medication guides, care points, and REMS
- NDC check and prescription verification processing
- third party adjudication, including major medical insurance and worker's comp
- accounts receivable processing
- point of sale capabilities
- DME/CMS 1500 on-line billing
- 340B inventory control management and reports
- wholesaler interface with automated ordering
- nursing home and LTC batch processing
- wireless signature capture/delivery manifest module
- MTM services billing electronically
- compounding services support

This technology was discussed in greater depth in *Chapter 6 Medication Order Entry and Fill Process*.

Barcoding technologies

Various studies have shown the potential for barcodes to reduce medication errors when dispensing medications from the pharmacy. Bedside barcode scanning, the use of barcode technology to verify a patient's identity, and the medication to be administered, is a promising strategy for preventing medication errors. Also, many community pharmacies will scan the barcodes on the manufacturer's bottles to verify that the same NDC number entered in the pharmacy's computer system matches the NDC number on the drug being dispensed. Its use has been increasing in notable places, such as the Veterans Affairs hospitals, University of Pittsburgh Medical Centers (UPMC), many of the major community pharmacy chains, and various hospitals throughout the United States.

Barcode medication verification at the bedside is usually implemented in conjunction with an electronic medication administration records (eMAR), allowing health care providers to automatically document the administration of drugs by means of barcode scanning. Because the eMAR imports medication orders electronically from either the computerized prescriber order entry (CPOE) or the pharmacy management software, its implementation may reduce transcription errors. Given its potential to improve medication safety, barcoding, in combination with eMAR technology, is being considered as a criterion for achieving "meaningful use" of health information technology and for obtaining financial incentives under the American Recovery and Reinvestment Act of 2009.

Automated counting and dispensing technologies

Counting out large numbers of tablets and capsules by hand can be a very laborious task, especially in today's high volume pharmacies. Tablet counting machines are able to assist in this process, as they can count tablets quickly and accurately. These machines will manage these rapid counts in a number of ways. Some will use an electronic 'eye' to count the medications as they fall through, others will know the weight of specific medication tablets or capsules and determine the quantity based on the total weight poured into the vial, while others will use specially designed hoppers that will deliver medications as the tray inside the cassette spins.

Many of the new high-end tablet counters can even receive prescriptions from the pharmacy management software, fill the vials, and label them once the ordered has been processed. Kirby Lester, McKesson Baker Cells, and ScriptPro Systems are commonly used automated tablet counters.

Automated compounding devices

An automated compounding device (ACD) is a machine used to prepare compounded sterile preparations, such as parenteral nutrition. The manual method of parenteral nutrition admixture compounding is labor-intensive and requires multiple manipulations of infusion containers, sets, syringes, needles, and so forth, which can lead to the extrinsic contamination of the final admixture with sterile and nonsterile contaminants.

These ACDs not only automate the compounding process providing improved turn around time and accuracy, but many of the newer ACDs can also help with many of the more complex concepts such as: limits for anions/cations, osmolarity warnings, volume warnings, and calcium/phosphate solubility.

Automated cart-fill systems

Cart-fill commonly involves pharmacies preparing a 24 hour supply for all the medications in a facility, and then delivering these medications to the nursing units into patient specific containers often located on carts (hence the name). Cart-fills can be a tedious operation for technicians to fill and for pharmacists to check. Robotic arms and automatic tablet dispensing and packaging systems provide options for automating this process.

Robotic arms have been developed for pharmacies that can operate on vertical and horizontal rails. Operating in an enclosed area, these robotic arms move up and down aisles filled with pegs containing individual drug doses stored in plastic, bar-coded pouches. The arms pick the appropriate medications and drops them into either patient specific envelopes or cassettes, which can then be used for the cart-fill.

Automatic tablet dispensing and packaging systems may vary in design and capacity, but they typically contain hundreds of containers with each container being capable of holding a different medication. The medications will be dispensed in long strips that contain patient specific medications in the order that the patient is scheduled to receive them.

Automated storage and retrieval systems

As the name implies, these systems may be used to both store and retrieve pharmacy inventory. Instead of needing to move up and down aisles in the pharmacy to acquire this inventory, these systems can store materials in tightly packed spaces on vertically rotating carousels that can bring the items to the staff. When an automated pick list is either run on the machine, or if a particular medication barcode is scanned, the machine will properly rotate the materials to the pharmacy team member.

Automated dispensing cabinets

An automated dispensing cabinet (ADC) is a computerized drug storage device or cabinet designed for institutional settings. ADCs allow medications to be stored and dispensed near the point of care, while controlling and tracking drug distribution. Facilities that use ADCs typically use them to manage floor stock medications, PRNs, and controlled substances. Some pharmacies will also use them to provide quick access to emergency medications, common drugs, or even all the medications on a patient profile.

Due to the fact that these systems are commonly located throughout a hospital, additional security is often implemented on these systems ranging from the use of user names and passwords, to barcode scanning and biometric identification.

Automated prescription pick-up

Automated prescription pick-up can have two potential pieces. The first, which is very widely utilized, are automatic prescription refill services where the pharmacy will automatically generate a refill for specific patient medications. Many of these systems will notify the patient through telephone message, email, or even text message that their refill is ready.

The second part, which is less common, is placing medications that have already been filled into machines that patients can access without going to the pharmacy counter. This frees up the staff from waiting on customers that do not need additional services, and can help reduce work loads during high volume periods of the day. It also provides the opportunity for a patient to pick-up a medication after the pharmacy has already closed. If a patient does decide they have a question for a pharmacist, these systems include phones and video cameras so that they can still receive counseling from either an on-site pharmacist or even one located remotely.

Reconciliation and profit analyzing technologies

A challenge for many pharmacies can be to receive full and proper payment for all their medications dispensed. Automated rebilling services can be set-up to receive orders that didn't pay-out, and look at what might be done differently to get the prescriptions to process. Also, these systems often have the ability to connect to the pharmacy software and correct prices related to average wholesale prices (AWP) for both brand and generic medications, which in-turn will also help with pharmacy revenues.

Pharmacy kiosks

Some facilities, particularly grocery stores without pharmacies, have started using pharmacy kiosks. These kiosks have the ability to scan patient insurance cards, IDs, and prescriptions for a pharmacy to remotely review and enter. These machines will contain the most common medications (usually excluding controlled substances), and with a pharmacist's authorization, will fill these prescriptions, including placing patient labels on the vials. Through the use of video cameras and phone receivers, a pharmacist will offer patient counseling at the kiosk. The patient will typically be able to pay for their prescription with either a credit card or debit card to complete the process.

Telemedicine

Telemedicine is the use of telecommunication and information technologies in order to eliminate distance barriers and improve access to medical services. Telemedicine can be broken into three main categories: store-and-forward, remote monitoring, and (real-time) interactive services.

Store-and-forward telemedicine involves acquiring medical data, and then transmitting this data to an appropriate medical professional at a convenient time for assessment off-line. It does not require the presence of both parties at the same time.

Remote monitoring, also known as self-monitoring or testing, enables medical professionals to monitor a patient remotely using various technological devices. This method is primarily used for managing chronic diseases or specific conditions, such as heart disease, diabetes mellitus, or asthma.

Interactive telemedicine services provide real-time interactions between patient and provider, to include phone conversations, online communication and home visits. Many activities such as history review, physical examination, psychiatric evaluations, and ophthalmology assessments can be conducted through this means.

Telepharmacy

Telepharmacy is the delivery of pharmaceutical care via telecommunications to patients in locations where they may not have direct contact with a pharmacist. It is an instance of the wider phenomenon of telemedicine, as implemented in the field of pharmacy. Telepharmacy services include drug therapy monitoring, patient counseling, prior authorization and refill authorization for prescription drugs, and monitoring of formulary compliance with the aid of teleconferencing or videoconferencing. Remote dispensing of medications by automated packaging and labeling systems can also be thought of as an instance of telepharmacy. Telepharmacy services can be delivered at retail pharmacy sites or through hospitals, nursing homes, or other medical care facilities.

The term can also refer to the use of video conferencing in pharmacy for other purposes, such as providing education, training, and management services to pharmacists and pharmacy staff remotely.

Pneumatic tube systems

Pneumatic tubes are systems that propel cylindrical containers through a network of tubes by compressed air or by partial vacuum. These systems are sometimes employed for pharmacy drive-throughs, to provide increased convenience to customers. These systems are even more common in hospital settings, providing a rapid means to transport medications, devices, and other necessary materials throughout the facility in a very rapid manner.

While pneumatic tubes can safely transport most items under 2 kg, there are some items that institutions and pharmacies ban including:

- controlled substances, if the tubes are not able to be sent and received in a sufficiently secure manner,
- drugs with long protein chains, as there is a concern that the vibrations in a pneumatic tube system may damage them,
- glass containers, also because of concern that the vibrations may damage them,
- live modalities (i.e., medicinal leeches),
- hazardous liquids due to concerns of leakage and exposure of the entire pneumatic tube system to these materials.

While pneumatic tube systems are not a new technology, they continue to improve and gain more complex features, such as having security codes that can be changed on the fly so only authorized individuals can access them, the ability to notify specific staff members when particular tubes arrive on the unit, and the inclusion of GPS tracking devices to identify their physical location.

Robotic delivery systems

If materials need to be delivered to the nursing units, and these items cannot be sent using a pneumatic tube system, another option involves the use of robotic delivery systems. These robots can navigate through hospital corridors and elevators to transport materials from the pharmacy to the hospital floors. These robots can work around the clock, and make both scheduled and on-demand deliveries. They can deliver something as small as a single IV or an entire unit's medication order.

Quick review question 13

Electronic health records often share information through what kind of system(s)?

 a. HIE
 b. IVR
 c. ACD
 d. all of the above

Quick review question 14

Which of the following is not an improvement associated with e-prescribing?

 a. improved patient safety
 b. reduces the probability that a prescriber will select the wrong item on a menu driven interface
 c. better formulary adherence
 d. reducing the time pharmacy spends calling the prescribers

Quick review question 15

Which of the following is not an improvement associated with CPOE?

 a. health system reporting
 b. patient safety
 c. reduces the probability that a prescriber will select the wrong item on a menu driven interface
 d. billing accuracy

Quick review question 16

What do pharmacies commonly use their IVR for?

 a. phone calls
 b. Patients will order refills through the IVR.
 c. Physicians frequently use IVRs to phone in new prescriptions and/or clarifications.
 d. all of the above

Quick review question 17

Which of the following technologies may have the ability to perform prescription processing, work flow management, offer medication guides, and provide third party adjudication?

a. interactive voice response
b. automated compounding devices
c. pharmacy management software systems
d. automated storage and retrieval systems

Quick review question 18

What can be verified by scanning the barcode on a manufacturer's bottle?

a. NDC
b. lot number
c. expiration date
d. It can verify the information written to the RFID chip beneath the label on the bottle.

Quick review question 19

How do automated counting and dispensing technologies actually count the quantity of medication being dispensed?

a. An electronic 'eye' counts the number of tablets that fall past it.
b. They calculate the number of tablets based on weight.
c. A specially designed hopper will dispense a specific number of tablets based on how many revolutions it makes.
d. all of the above

Quick review question 20

Cart-fill is traditionally considered a labor intensive task that provides many opportunities for dispensing errors. Which of the following technologies can be used to fully automate this process.

a. CPOE
b. robotic arms
c. pharmacy kiosks
d. automated storage and retrieval systems

Quick review question 21

Which of the following is not a true statement about automated storage and retrieval systems?

a. They decrease the amount of space you would otherwise need to store these products.

  b. They reduce the distance staff need to move to acquire inventory.

  c. They reduce the costs of the medications.

  d. They reduce the amount of time required to gather inventory.

Quick review question 22

Which technology is designed to store and dispense medications near the point of care in institutional settings?

  a. automated counting devices
  b. automated compounding devices
  c. automated dispensing cabinets (a.k.a., medstations)
  d. there is no technology designed to fill that niche

Quick review question 23

Which of the following terms are best defined as, "The use of telecommunication and information technologies in order to eliminate distance barriers and improve access to medical services."?

  a. telemedicine
  b. interactive voice response systems
  c. pharmacy kiosks
  d. automated storage and retrieval systems

Quick review question 24

Which of the following items should not be sent from the pharmacy to a nursing unit using a pneumatic tube system?

  a. a 50 mL minibag with 32.5 mg of doxorubicin
  b. a missing dose of terazosin 5 mg capsules
  c. a 100 mL minibag with 3 g of Unasyn (ampicillin and sulbactam)
  d. a stat dose of aspirin

Quick review question 25

Which of the following medications would you not expect to be dispensed by a pharmacy kiosk?

  a. metoprolol tartrate
  b. controlled release oxycodone

c. tamoxifen citrate
d. esomeprazole

Quick review question 26

Why might a pharmacy want to set-up an automated prescription pick-up machine?

a. to help patients receive faster service
b. to lighten the burden on the pharmacy staff during busy periods of time
c. to provide a means for customers to receive their medication after the pharmacy is already closed
d. all of the above

Quick review question 27

Telemedicine is typically broken into three categories: store-and-forward, remote monitoring, and (real-time) interactive services. Which category would off-site monitoring of blood sugar fall into if a glucometer periodically links readings with time stamps back to their profile for the health care team to review?

a. store-and-forward
b. remote monitoring
c. interactive services
d. none of the above

Quick review question 28

What kind of automation technology typically contain hundreds of containers, with each container being capable of holding a different medication? The medications in it will be dispensed in long strips that contain patient specific medications in the order that the patient is scheduled to receive them.

a. automatic tablet dispensing and packaging systems
b. automated dispensing cabinets (a.k.a., medstations)
c. telemedicine
d. HIE

Quick review question 29

Telepharmacy is considered a specialized area of what other technology?

a. telemedicine
b. electronic health records
c. interactive voice response systems
d. health information exchanges

Challenges to implementing pharmacy automation and technology

While it is common to focus on the positives associated with pharmacy automation and technology, there are significant challenges to overcome including: staff and customers that are change adverse, direct and indirect costs associated with implementing new automation and technology, and concerns over maintaining patient confidentiality.

Resistance to change

Most people achieve certain routines in how they perform various tasks, whether they are simple or complex, and in pharmacy this is true of the pharmacy staff as well as their clients (patients, nurses, physicians, etc.). They have certain expectations as to how long a particular task will take and what the outcome will be. A challenge to implementing any type of automation or technology is that it will require changes to these routines in order to be implemented. Generally, it will help if all individuals involved know what the new technology is, why it is being implemented, and discussion about what the expected gains are versus the potential challenges can be reviewed. It is also helpful to follow-up after implementation to discuss any unforeseen outcomes, whether good or bad.

Cost

While these automations claim they can reduce existing HR salaries and benefits, improve productivity and workflow efficiency, and enhance operation services, there is the challenge that the costs on some of these technologies can be overwhelming. Depending on complexity, staff knowledge, product competition, and sheer manufacturing costs, the prices on this equipment can range from free (if looking at an open source IVR that your own staff knows how to install and maintain, although staff time does till have some costs associated with it) to millions of dollars (such as some of the high end robotic arms that perform cart-fill procedures and fill first-dose medications in some institutions). Another cost to consider is the support contracts that pharmacies often need for all this cutting edge technology.

Confidentiality issues

While all this health information collection and integration provides opportunities to improve patient care, it does raise concerns over maintaining patient confidentiality, as so many health care entities may have access. According to a *Los Angeles Times* article (*At Risk of Exposure*, June, 6, 2006), roughly 150 people (from doctors and nurses, to technicians and billing clerks) have access to at least part of a patient's records during a hospitalization, and 600,000 payers, providers, and other entities that handle providers' billing data have some access as well.

There are three potential sources for security failures with respect to this information: human, environmental threats, and technology failures. Human threats include employees inappropriately accessing information and/or not securing it, as well as threats from computer crackers (sometimes called hackers by the media). Environmental threats can range from earthquakes and hurricanes to fires and floods. These environmental threats may unintentionally expose patient sensitive information, commonly referred to as protect health information (PHI). Technology failures could include hardware and software failures that could either be related to the other two security failures or independent of them.

Standards need to be met to ensure the security of this information, along with strict guidelines about who should have access to specific patient information. The Health Insurance Portability and Accountability Act of 1996 does publish standards for security. The security standards are designed to protect the confidentiality of PHI that is threatened by the possibility of unauthorized access and interception during electronic transmission. Like the privacy provisions, any pharmacy that transmits any health information in electronic form is required to comply with the security rules.

In particular, the security standards define administrative, physical, and technical safeguards that the pharmacist must consider in order to protect the confidentiality, integrity, and availability of PHI.

A unique aspect of the security provisions is that they include both "required and addressable" implementation specifications. Required implementation specifications are those that must be met, whereas, in addressable specifications, the pharmacy must determine whether the suggested safeguards are reasonable and appropriate, given the size and capability of the organization, as well as the risk.

While cost may be a factor that a covered entity may consider in determining whether to implement a particular specification, nonetheless a clear requirement exists that adequate security measures be implemented. Cost considerations are not meant to exempt covered entities from this responsibility.

Quick review question 30

Which of the following is a potential source for a security risk with regards to patient specific health information being electronically transmitted?

  a. human
  b. environmental threats
  c. technology failures
  d. all of the above

Quick review question 31

Which law provides standards for security of protected health information?

  a. Resource Conservation and Recovery Act
  b. Health Insurance Portability and Accountability Act
  c. Patriot Act
  d. Medicare Modernization Act

The future of pharmacy information systems

With all of these advancements, the true end-goals are improved patient outcomes through a variety of means including: health care providers with greater access to the relevant information, more educated patients, improved medication and therapy compliance, better reporting, reduced errors, and more opportunities for patients to communicate with their health care providers. While great advancements have been made in these areas, more integration needs to be achieved as these technologies mature and new ones are developed.

As these services and information systems become more advanced, they are changing the role of both pharmacists and pharmacy technicians. Some particular areas of expected growth include:

- As electronic health records eventually achieve full integration with a patient's entire medical history, it will allow for better screening of potential medication errors, and help to better identify patients that may benefit from from medication therapy management (MTM).
- Greater autonomy will likely be achieved by pharmacy technicians as they can rely on automation to reduce the possibilities of making errors.
- Pharmacists and pharmacy technicians will need to learn more about how to properly interact with and maintain this equipment.
- This specialized area of knowledge will also offer future growth opportunities to pharmacists and pharmacy technicians that understand both pharmacy and the information technology component. These specialized pharmacy personnel will be utilized by pharmacies, institutions, and the automation companies.

Quick review question 32

As pharmacy automation and technology continue to advance, which of the following is a likely outcome?

a. The responsibilities of pharmacists and pharmacy technicians will decrease as they can let technology do all their work.

b. Humans will no longer need to seek employment in health care as our electronic over lords will take care of us.

c. There will be more specialized positions for pharmacists and pharmacy technicians in the field of pharmacoinformatics.

d. There will likely be more limitations as to what pharmacy personnel will be allowed to do, and an increase of responsibility will be shifted to other health care professionals.

Answers to quick review questions

1. D - Pharmacy informatics (a.k.a., pharmacoinformatics) is the use and integration of data, information, knowledge, technology, and automation in the medication use process.
2. A - A computer cracker is someone who makes unauthorized use of a computer, especially to tamper with data or programs. The media commonly refers to these individuals as hackers.
3. B - A database management system (DBMS) is a software system designed to allow the definition, creation, querying, update, and administration of databases. Some common examples of DBMSs include MySQL, MariaDB, PostgreSQL, SQLite, MS SQL Server, Microsoft Access, Oracle Database, SAP, dBASE, and IBM DB2.
4. D - Databases are able to readily organize large quantities of data that can be easily queried. This information is quickly retrievable, and databases do not tire of repetitive search strings, achieving a higher degree of accuracy and speed than what could be achieved without them.
5. C - A common challenge when using DBMSs is getting the information from one database to communicate with a different database.
6. B - Database users can be limited to specific data subsets called "subschemas".
7. D - Pharmacy information systems have become so common and well integrated that they can directly impact every step of the medication use process.
8. A - It has become a common practice for prescribers to use e-prescription software to order outpatient prescriptions.

9. B - It has become a common practice for prescribers to use computerized prescriber order entry software to order inpatient prescriptions.
10. B - Bedside barcode scanners can be utilized to assure that the right drug is being given to the right patient at the right time, and even ensure that it provides the right documentation by automatically charting when it is being given; but bedside barcode scanners cannot provide assurance that a care giver has the right attitude.
11. D - Remote monitoring can be used to track blood pressure, blood glucose, and medication adherence.
12. C - Pharmacies commonly use interactive voice response (IVR) systems for direct phone calls, order refills, and physicians will frequently use them to phone in new prescriptions and/or clarifications.
13. A - Health information exchanges (HIEs) help to share information across different health care settings in electronic format.
14. B - While there are many advantages/improvements associated with the use of e-prescribing, there is a real possibility of prescribers making fat finger errors, especially if they are rushed, distracted, or just having difficulty making proper selections from crowded screens.
15. C - While there are many advantages/improvements associated with the use of CPOE, there is a real possibility of prescribers making fat finger errors, especially if they are rushed, distracted, or just having difficulty making proper selections from crowded screens.
16. D - Pharmacies commonly use interactive voice response (IVR) systems for direct phone calls, order refills, and physicians will frequently use them to phone in new prescriptions and/or clarifications.
17. C - Pharmacy management systems are usually capable of performing a wide range of tasks including: prescription processing, e-prescribing, prescription pick-up with signature capture, refill queue, work flow manager, prescription scanning and patient card scanning, medication guides, care points, and REMS, NDC check and prescription verification processing, third party adjudication including major medical insurance and worker's comp, accounts receivable

processing, point of sale capabilities, DME/CMS 1500 on-line billing, 340B inventory control management and reports, wholesaler interface with automated ordering, nursing home and LTC batch processing, wireless signature capture/delivery manifest module, MTM services billing electronically, and compounding services support.

18. A - Currently, the barcode on a manufacturer's bottle only contains the NDC.

19. D - Tablet counting machines manage pill counts in a number of ways. Some will use an electronic 'eye' to count the medications as they fall through, others will know the weight of specific medication tablets or capsules and determine the quantity based on the total weight poured into the vial, while others will use specially designed hoppers that will deliver medications as the tray inside the cassette spins.

20. B - Robotic arms have been developed for pharmacies that can operate on vertical and horizontal rails. Operating in an enclosed area, these robotic arms move up and down aisles filled with pegs containing individual drug doses stored in plastic, bar-coded pouches. The arms pick the appropriate medications, and drops them into either patient specific envelopes or cassettes, which can then be used for the cart-fill.

21. C - While automated storage and retrieval systems can provide a variety of tangible benefits, they do not reduce the cost of the inventory that the pharmacy orders.

22. C - ADCs allow medications to be stored and dispensed near the point of care while controlling and tracking drug distribution.

23. A - Telemedicine is the use of telecommunication and information technologies in order to eliminate distance barriers and improve access to medical services.

24. A - Doxorubicin is considered a hazardous drug (see *CHAPTER 1 Pharmacology for Technicians*), and hazardous liquids should not be sent in pneumatic tube systems due to concerns of leakage and exposure of the entire pneumatic tube system to these materials.

25. B - Pharmacy kiosks will contain the most common medications, usually excluding controlled substances.

26. D - An automated prescription pick-up machine frees up the

staff from waiting on customers that do not need additional services and can help reduce work loads during high volume periods of the day. It also provides the opportunity for a patient to pick-up a medication after the pharmacy has already closed.

27. B - Remote monitoring, also known as self-monitoring or testing, enables medical professionals to monitor a patient remotely using various technological devices. This method is primarily used for managing chronic diseases or specific conditions, such as heart disease, diabetes mellitus, or asthma.

28. A - Automatic tablet dispensing and packaging systems may vary in design and capacity, but they typically contain hundreds of containers with each container being capable of holding a different medication. The medications will be dispensed in long strips that contain patient specific medications in the order that the patient is scheduled to receive them.

29. A - Telepharmacy is the delivery of pharmaceutical care via telecommunications to patients in locations where they may not have direct contact with a pharmacist. It is an instance of the wider phenomenon of telemedicine, as implemented in the field of pharmacy.

30. D - There are three potential sources for security failures with respect to this information: human, environmental threats, and technology failures.

31. B - The Health Insurance Portability and Accountability Act of 1996 does publish standards for security designed to protect the confidentiality of PHI that is threatened by the possibility of unauthorized access.

32. C - As pharmacy automation and technology continue to advance, there will be growth opportunities for pharmacists and pharmacy technicians that understand both pharmacy and the information technology component.

CHAPTER 10 Practice Exams for the Pharmacy Technician Certification Exam

This chapter is intended to provide two complete practice exams with all categories of questions proportionally represented to how they will appear on the Pharmacy Technician Certification Exam. The answers for each practice exam are available immediately after the exam. If this were the actual exam, you would have a 1 hour 50 minute time limit, and require a minimum score of 650 out of 900 (approximately a 72%) to pass. With that in mind, these practice exams should give you a good opportunity to gage how well prepared you are for the actual exam. With each question, choose the best available answer.

Practice exam 1

1. If your pharmacy stocks 20 milliliter vials of 1% lidocaine, how many milliliters would you need to draw up to fulfill a request for 100 mg of lidocaine?

    a. 5 mL
    b. 10 mL
    c. 20 mL
    d. 100 mL

2. What is the BSA of a 6 foot tall patient that weighs 196 pounds?

    a. 1.73 m^2
    b. 1.98 m^2

c. 2.13 m^2
d. 4.53 m^2

3. Which of the following statements best defines parenteral administration?

   a. a route of administration requiring permission from an adult or guardian
   b. a route of administration other than the GI tract
   c. an oral route of administration
   d. a route of administration involving the rectum

4. Which of the following penicillin derivatives with beta-lactamase inhibitors is available in an oral formulation?

   a. amoxicillin & clavulanate
   b. ampicillin & sulbactam
   c. piperacillin & tazobactam
   d. ticarcillin & clavulanate

5. Which of the following medications carries a black box warning about increased risk of tendon rupture?

   a. Avelox
   b. Augmentin
   c. Zithromax
   d. Vibramycin

6. Which one of the following items listed is not likely to interact with warfarin?

   a. aspirin
   b. vitamin K
   c. grapefruit juice
   d. broccoli

7. What is the proper term used for when a pharmacy charges for a medication that the patient did not receive?

a. fraud
b. human error
c. profit
d. insufficient data to properly answer this question

8. A common reject code for Medicare patients that have a prescription for Fioricet (butalbital, acetaminophen, and caffeine) is '70'. What does this reject code mean?

   a. product/service not covered
   b. refill too soon
   c. missing/invalid prescriber ID
   d. switch vendor is currently down

9. Which organization is responsible for providing medical and public health services to members of federally recognized Tribes and Alaska Natives?

   a. CHAMPVA
   b. TRICARE
   c. IHS
   d. SCHIP

10. Which of the following is an example of quality assurance?

   a. Carvedilol was accidentally dispensed incorrectly from the pharmacy instead of captopril. The root cause was because the technician accidentally grabbed the wrong unit dose package, as they are both small white tablets that are close to each other in the alphabet. The solution was to create a new policy that each drug will be set physically separate from each other on different shelves.
   b. checking a floor stock list to ensure it has been filled correctly before it leaves the pharmacy
   c. The pharmacy devises a method to make sure that parenteral nutrition solutions above 1000 mOsmol/L are never infused in a peripheral IV line.
   d. having a policy that all chemotherapeutic agents must be checked by at least two pharmacists prior to dispensing

11. What are drug recalls based on particular batches identified by?

    a. expiration date
    b. lot number
    c. NDC
    d. schedule

12. Electronic health records often share information through what kind of system?

    a. ADC
    b. HIE
    c. VOIP
    d. magic

13. There are three potential sources for security failures with respect to keeping PHI secure: human, environmental threats, and technology failures. Which of the following is an example of a human threat?

    a. employees inappropriately accessing information
    b. employees not adequately securing patient data
    c. potential threats from computer crackers
    d. all of the above

14. Which prescription origin code (POC) should be used if a physician sends an order to a pharmacy via a fax machine?

    a. 1
    b. 2
    c. 3
    d. 4

15. A prescription is written for E.E.S. 200 mg/5 mL, Disp 200 mL, Sig: ss Tbs po tid ac for 7 days, Refills: NR. How much medication should the patient take for one dose?

    a. one-half teaspoonful
    b. one-half tablespoonful

c. one teaspoonful
d. one tablespoonful

16. The majority of medications need to be stored at controlled room temperature. What is controlled room temperature according to the USP?

   a. -25° to -10° C
   b. 2° to 8° C
   c. 15° to 30° C
   d. 59° to 86° C

17. A prescription is written for Wellbutrin 100 mg tabs, Disp: #90, Sig: i tab po tid, Refills: 5. How should this prescription be adjusted to provide the medication in 30 day increments?

   a. dispense 30 tabs per fill with 11 refills and no partial fills
   b. dispense 60 tabs per fill with 8 refills and no partial fills
   c. dispense 90 tabs per fill with 5 refills and no partial fills
   d. dispense 120 tabs per fill with 3 refills and a partial fill of 60 tablets

18. What are the special storage requirements for acetazolamide, doxycycline, zolmitriptan, and linezolid?

   a. refrigeration
   b. non-PVC containers
   c. light resistant packaging
   d. oral syringes

19. How many bottles of fluoxetine 20 mg capsules should be ordered based on the following information?
Package size: 100 capsules/bottle
Minimum number of units: 2000 capsules
Maximum number of units: 6000 capsules
Current inventory level: 1870 capsules

   a. 0 bottles
   b. 4 bottles
   c. 41 bottles

d. 60 bottles

20. If a pharmacy is using direct purchasing for a particular medication, who is the pharmacy purchasing the medication from?

a. the wholesaler
b. the manufacturer
c. another pharmacy
d. the patient

21. Which of the following groups of medications do not reduce inflammation?

a. opioids (oxycodone, codeine, etc.)
b. NSAIDs (ibuprofen, naproxen, etc.)
c. salicylates (aspirin, mesalamine, etc.)
d. glucocorticosteroids (prednisone, methylprednisolone, etc.)

22. Hospital acquired pneumonia is an example of what kind of infection?

a. community
b. koinos
c. nosocomial
d. none of the above

23. Which of the following is an example of an inscription on a prescription order?

a. Lipitor 20 mg
b. Disp: #30
c. i tab PO daily
d. the physician's name signed above a line stating, "Product Selection Permitted"

24. How does an individual acquire a private health insurance policy?

a. through their employer
b. through a family member's employer
c. through individual purchase from an insurance company or from an exchange
d. all of the above

25. Which of the following is the best practice for requesting the following digoxin prescription?

a. digoxin 250 µg tab, Disp 30, Sig: i tab po daily
b. digoxin .25 mg tab, Disp 30, Sig: 1 tab po daily
c. digoxin 250 mcg tab, Disp 30, Sig: 1 tab po qd
d. none of the examples follow best practices

26. If a customer comes to the pharmacy counter and asks for advice about a side effect related to their medication, what should the technician do?

a. the technician should give advice
b. have the pharmacist counsel the patient
c. the patient should be told to contact their PCP with such questions
d. ask the patient if they have access to the internet

27. Which of the following classes is not considered a high alert medication group?

a. narcotics/opioids
b. NSAIDs
c. parenteral nutrition
d. antithrombotic agents

28. Which Act mandates the use of black bins for environmentally hazardous waste?

a. PPPA
b. RCRA
c. HIPAA
d. TJC

29. You receive a prescription for a controlled substance written by a Julius Erving, M.D., and it has the following DEA number recorded on it - AE65432130. Does Dr. Erving's DEA number validate, and if not, then why does it not validate?

    a. this DEA number validates
    b. the letter(s) at the beginning of this DEA number are incorrect
    c. incorrect check sum
    d. incorrect number of digits

30. Which of the following would not be considered PHI?

    a. Brea Larkin, prescribed escitalopram 10 mg daily
    b. 1650 Metropolitan Street, 54 y.o. woman, shingles
    c. SSN 645-78-1209, cellulitis on right inner thigh
    d. 10% of patients receiving metronidazole had a side effect of diarrhea

31. If a pharmacy compounds and dispenses 30 grams of a 2.5% hydrocortisone cream made from hydrocortisone powder and Eucerin cream, how much hydrocortisone should have been used to compound it?

    a. 0.75 mg
    b. 2.5 g
    c. 30 g
    d. 750 mg

32. Which syringe size would be most appropriate to use if you required 3.2 mL of solution?

    a. tuberculin syringe
    b. 3 cc syringe
    c. 5 cc syringe
    d. 10 cc syringe

33. Which of the following medications should women that are pregnant or are trying to become pregnant avoid handling if

the tablets are crushed or broken?

a. doxazosin
b. ephedrine
c. finasteride
d. labetalol

34. Which of the following diuretics is considered potassium sparing?

a. Bumex
b. Lasix
c. Microzide
d. Aldactone

35. If a Medicare patient gets a prescription for metformin filled, which part of their insurance should it be processed under?

a. Medicare Part A
b. Medicare Part B
c. Medicare Part D
d. The patient will need to pay the cash price.

36. What kind of reimbursement plan would a capitation fee be classified as?

a. fee-for-service reimbursement
b. episode-of-care reimbursement
c. oikos nomos
d. none of the above

37. When donning PPE to prepare sterile products, in what order should you don your PPE?

a. from top to bottom
b. from cleanest to dirtiest
c. from dirtiest to cleanest
d. the order does not matter

38. Which of the following is an example of a pharmacy

technology that can be utilized by a physician to prescribe medications?

    a. e-prescribing
    b. CPOE
    c. IVR
    d. all of the above

39. Which of the following is a potential barrier to implementing pharmacy automation and technology?

    a. staff and customers that are change adverse
    b. direct and indirect costs associated with implementing new automation and technology
    c. concerns over maintaining patient confidentiality
    d. all of the above

40. A prescription is written for Timoptic 0.5%, Disp: 15 mL, Sig: i gtt o.s. bid, Refills: 2. What is the route of administration in this prescription?

    a. left eye
    b. right eye
    c. left ear
    d. right ear

41. A prescription is written for Lantus SoloStar, Disp: 2 boxes (15 mL/box), Sig 40 U SQ q pm, Refills 5. How should this prescription be adjusted if the patient prefers a 90 day supply in order to get three months worth of medication for only a two month supply copay (a savings offered through his insurance)?

    a. dispense 3 boxes per fill with 3 refills and no partial fills
    b. dispense 4 boxes per fill with 2 refills and no partial fills
    c. dispense 2 boxes per fill with 5 refills and no partial fills
    d. dispense 1 boxes per fill with 11 refills and no partial fills

42. A prescription is written for quinine 324 mg #42 2 caps po q8h. How many days' supply does this prescription allow

for?

    a. 7
    b. 14
    c. 21
    d. 42

43. How many bottles of metoprolol tartrate 50 mg tablets
    should be ordered based on the following information?
    Package size: 100 tablets/bottle
    Minimum number of units: 120 tablets
    Maximum number of units: 480 tablets
    Current inventory level: 90 tablets

    a. 0 bottles
    b. 1 bottle
    c. 2 bottles
    d. 3 bottles

44. How many bottles of tramadol 50 mg tablets should be
    ordered based on the following information?
    Package size: 100 tablets/bottle
    Minimum number of units: 200 tablets
    Maximum number of units: 800 tablets
    Current inventory level: 150 tablets

    a. 0 bottles
    b. 1 bottle
    c. 6 bottles
    d. 7 bottles

45. The Greek letter 'μ' is an abbreviation that is supposed to be
    avoided because it can be misinterpreted. If a prescription
    arrives in the pharmacy noting a medication strength in μg,
    how should it be interpreted?

    a. micrograms
    b. milligrams
    c. Megagrams
    d. units

46. If a pharmacy transfers a prescription to your pharmacy, and you are not sure how they originally received it, which POC should be marked?

    a. 0
    b. 1
    c. 2
    d. 3

47. Which of the following is a high-alert medication?

    a. metronidazole
    b. normal saline solution
    c. D5W
    d. heparin

48. The MERP reporting program is managed by which organization?

    a. CDC
    b. FDA
    c. ISMP
    d. TJC

49. What is the federal controlled substance schedule of Soma?

    a. CIII
    b. CIV
    c. CV
    d. on the federal level, this medication is not considered a controlled substance

50. The Omnibus Budget Reconciliation Act of 1990 imposed ProDUR requirements on pharmacists for Medicaid patients. What does ProDUR stand for?

    a. Professional Drug Use Regimens
    b. Peer Reviewed Operational Directives and Universal Rights

c. Prescription Drug Usage Records
d. Prospective Drug Utilization Review

51. A patient weighing 220 lbs is ordered zidovudine 1 mg/kg. How many milligrams of zidovudine should the patient receive?

    a. 100 mg
    b. 220 mg
    c. 1000 mg
    d. 2200 mg

52. What is the generic name for Unasyn?

    a. amoxicillin & clavulanate
    b. ampicillin & sulbactam
    c. piperacillin & tazobactam
    d. ticarcillin & clavulanate

53. Which of the following medications used to treat osteoporosis is not a bisphosphonate derivative?

    a. Reclast
    b. Boniva
    c. Forteo
    d. Fosamax

54. Which of the following situations is not an example of pharmacy communication?

    a. maintaining a personal list of BINs for the various insurances you process
    b. a staff meeting where proper use of opioids for pain management are discussed
    c. flagging a drug in the computer to let others know it has been recalled
    d. sending out an email to inform the staff about a change in call-off procedures

55. Which of the following is a common database interaction?

a. data definition and the ability to update said data
b. information retrieval
c. system administration
d. all of the above

56. Telemedicine is typically broken into three categories: store-and-forward, remote monitoring, and (real-time) interactive services. As telemedicine continues to expand, we are seeing an increase in services being offered that include video conferencing between the patient and various health care team members. Which of the aforementioned categories does this type of telemedicine belong in?

a. store-and-forward
b. remote monitoring
c. interactive services
d. this kind of interaction defies classification

57. A prescription is written for Novolin N insulin 10 mL, Disp: 1 vial, Sig: 20 U SQ q am and 25 U q pm. How many days' supply does this prescription allow for?

a. 2
b. 22
c. 30
d. 40

58. A prescription is written for DexPak 13 day TaperPak take u.d. How many days' supply does this prescription allow for?

a. 6
b. 13
c. 14
d. 30

59. Which of the following are OSHA required notices on hazardous substances?

a. MSDS

b. EOB
c. NIOSH
d. PPE

60. The use of PPE in an institutional setting is intended to prevent the spread of germs between the staff and whom or what?

a. patients
b. other staff members
c. sterile products
d. all of the above

61. What does it mean if you compare two products in the Orange Book and get a TE Code of "A"?

a. The two medications are considered therapeutic equivalents, and can be safely interchanged with each other.
b. There is a therapeutic equivalence problem.
c. The generic product has a grade range of 90-100%.
d. There is no such thing as a TE Code of "A".

62. Which of the following is not an example of a medication error?

a. giving a patient a medication intended for someone else
b. giving a medication intended for otic use, ophthalmically instead
c. giving a patient cyclobenzaprine 10 mg tablets when the physician ordered Flexeril 10 mg tablets
d. All of the above examples should be considered medication errors.

63. An adverse event to which of the following should be reported to VAERS?

a. cranberry pills
b. tadalafil
c. thiamine
d. Gardasil

64. You receive a prescription for a controlled substance written by a Ralph Quack, M.D., and it has the following DEA number recorded on it - BQ4811623. Does Dr. Quack's DEA number validate, and if not then why does it not validate?

    a. This DEA number validates.
    b. The letter(s) at the beginning of this DEA number are incorrect.
    c. incorrect check sum
    d. incorrect number of digits

65. According to USP 797, when should PECs be cleaned?

    a. at the beginning of each shift, and prior to each batch preparation
    b. every 30 minutes, during continuous compounding periods of individual CSPs
    c. when there are spills, and when surface contamination is known or suspected
    d. all of the above

66. Which of the hemopoietic agents listed below will stimulate the stem cells located in the bone marrow to produce more white blood cells?

    a. Aranesp
    b. Epogen
    c. Neupogen
    d. Neumega

67. Which of the following is an example of quality control?

    a. Carvedilol was accidentally dispensed incorrectly from the pharmacy instead of captopril. The root cause was because the technician accidentally grabbed the wrong unit dose package, as they are both small white tablets that are close to each other in the alphabet. The solution was to create a new policy that each drug will be set physically separate from each other on different shelves.

b. checking a floor stock list to ensure it has been filled correctly before it leaves the pharmacy

c. The pharmacy devises a method to make sure that parenteral nutrition solutions above 1000 mOsmol/L are never infused in a peripheral IV line.

d. having a policy that all chemotherapeutic agents must be checked by at least two pharmacists prior to dispensing

68. Some patients will order their refills from the pharmacy on the phone system by simply keying in the prescription number on the bottle when they reach the appropriate prompt. Which pharmacy technology is this an example of?

a. e-prescribing
b. telepharmacy
c. IVR
d. CPOE

69. How many boxes of albuterol inhalation solution vials for nebulization should be ordered based on the following information?
Package size: 25 vials/box
Minimum number of units: 50 vials
Maximum number of units: 150 vials
Current inventory level: 25 vials

a. 0 boxes
b. 1 box
c. 4 boxes
d. 5 boxes

70. A prescription is written for Duragesic 50 mcg/hr, Disp: 10 patches, Sig: 1 patch q72h. How many days' supply does this prescription allow for?

a. 3
b. 10
c. 30
d. 72

71. How many bottles of quetiapine 200 mg tablets should be ordered based on the following information?
    Package size: 100 tablets/bottle
    Minimum number of units: 200 tablets
    Maximum number of units: 800 tablets
    Current inventory level: 100 tablets

    a. 0 bottles
    b. 1 bottle
    c. 7 bottles
    d. 8 bottles

72. What kind of medication formulary is a managed care organization most likely to have?

    a. closed formulary
    b. open formulary
    c. proprietary formulary
    d. generic formulary

73. Placing valacyclovir and valganciclovir on different shelving units is an example of which kind of strategy to reduce medication errors?

    a. tallman lettering
    b. risk evaluation and mitigation strategy
    c. barcoding technology
    d. separating inventory

74. Which organization is responsible for listing drug recalls publicly?

    a. FDA
    b. NIOSH
    c. MSDS
    d. BOP

75. Which of the following medications is not a proton pump inhibitor?

a. metronidazole
b. omeprazole
c. rabeprazole
d. dexlansoprazole

76. Which of the following is a potential benefit from the improved efficiencies that often result from pharmacies adopting automation and establishing procedural routines?

    a. More time is available to provide counseling and MTM for the patients if they desire it.
    b. The patients often don't need to wait as long for prescriptions to be filled.
    c. Pharmacies may be capable of offering additional services as a result of less time being required by other processes.
    d. all of the above

77. A prescription is written for Sinemet 10/100, Disp: 180, Sig: ii tab po tid. How many days' supply does this prescription allow for?

    a. 30
    b. 60
    c. 90
    d. 180

78. Why do patients typically need a nitrate free interval?

    a. If the medication is taken routinely, it will cause a paradoxical effect.
    b. to minimize tolerance
    c. to help patients sleep at night
    d. There is no need for a nitrate free interval.

79. Labeling the shelves where Lamictal and Lamisil are kept as LaMICtal and LamISIL is an example of which kind of strategy to reduce medication errors?

    a. tallman lettering

b. risk evaluation and mitigation strategy
c. barcoding technology
d. separating inventory

80. Which of the following is a method to reduce particle size (comminution)? It may also include the grinding together of two or more substances in a mortar to mix them.

a. trituration
b. spatulation
c. micturation
d. geometric incorporation

81. If a Medicare patient is hospitalized, which part of their insurance should their hospital care be processed under?

a. Medicare Part A
b. Medicare Part B
c. Medicare Part D
d. The patient will need to pay the cash price.

82. If a pharmacy maintains hard copies of their prescriptions, what should CII prescriptions be marked with according to federal law before they are filed?

a. a red colored 'C'
b. a red colored 'CII'
c. a red colored 'N'
d. There are no federal regulations requiring that CII prescriptions be marked prior to filing.

83. Which NCC MERP classification does the following situation belong in? A diabetic patient with a severe sulfonamide allergy received a dose of glipizide, and had an allergic reaction requiring life sustaining measures. The patient eventually recovered, and was discharged from the hospital.

a. no error
b. error, no harm
c. error, harm

d. error, death

84. Which of the following medications is not considered a controlled substance?

    a. Luminal
    b. Fiorinal
    c. Fioricet
    d. Soma

85. What can be verified by scanning the barcode on a manufacturer's bottle?

    a. manufacturer
    b. strength, dosage form, and formulation
    c. package forms and sizes
    d. all of the above

86. Which of the following is the proper definition for the term 'osmolarity'?

    a. An insoluble substance separated from a solution due to a reaction between incompatible substances.
    b. A characteristic of a solution determined by the number of dissolved particles in it.
    c. The resulting solution when a drug is added to a parenteral solution.
    d. The process by which molecules intermingle as a result of their kinetic energy of random motion.

87. Which insurance programs are administered either in part or whole by CMS?

    a. Medicare
    b. Medicaid
    c. SCHIP
    d. all of the above

88. Under federal regulations, how many times may a prescription for schedule III-V medications be transferred?

a. none
b. once
c. five times, but only once per month
d. as often as a patient requests

89. On a prescription, which part is considered the signatura?

a. the directions to the patient
b. the medication ordered
c. the quantity to dispense
d. the prescriber's signature

90. If a medication is listed as expiring 02/2020, on what day does it expire?

a. January 31, 2020
b. February 1, 2020
c. February 28, 2020
d. February 29, 2020

Answers to practice exam 1

1. B	17.C	33.C
2. C	18.C	34.D
3. B	19.C	35.C
4. A	20.B	36.B
5. A	21.A	37.C
6. C	22.C	38.D
7. A	23.A	39.D
8. A	24.D	40.A
9. C	25.D	41.A
10.D	26.B	42.A
11.B	27.B	43.D
12.B	28.B	44.C
13.D	29.D	45.A
14.D	30.D	46.A
15.B	31.D	47.D
16.C	32.C	48.C

49.B	63.D	77.A
50.D	64.A	78.B
51.A	65.D	79.A
52.B	66.C	80.A
53.C	67.B	81.A
54.A	68.C	82.D
55.D	69.D	83.C
56.C	70.C	84.C
57.B	71.C	85.D
58.B	72.A	86.B
59.A	73.D	87.D
60.D	74.A	88.B
61.A	75.A	89.A
62.C	76.D	90.D

Practice exam 2

1. Which regulation requires pharmacies to make a reasonable attempt to provide privacy for patients receiving counseling?

   a. Medicare Modernization Act
   b. OBRA-90
   c. HIPAA
   d. PHI

2. If a medication functions as a plasticizer, which of the following materials may be appropriate to use for storing the medication in?

   a. PVC containers
   b. glass containers
   c. both a and b
   d. none of the above

3. A prescription is written for Z-Pak 500 mg PO on first day of therapy, then 250 mg PO once daily for the next 4 days. Total cumulative dose 1.5g. How many days' supply does this prescription allow for?

a. 4
b. 5
c. 7
d. 10

4. Which classes of medications have significant concerns with respect to grapefruit juice drug interactions (GJDI)

   a. calcium channel blockers
   b. HMG-CoA reductase inhibitors
   c. both a and b
   d. Medications are not affected by grapefruit juice.

5. What kind of storage is required for volatile or flammable substances (including tax free alcohol)?

   a. a cool location
   b. a properly ventilated location
   c. a storage area designed to reduce fire and explosion potential
   d. all of the above

6. Which of the following is the proper definition for the term 'precipitate'?

   a. An insoluble substance separated from a solution due to a reaction between incompatible substances.
   b. A characteristic of a solution determined by the number of dissolved particles in it.
   c. The resulting solution when a drug is added to a parenteral solution.
   d. The process by which molecules intermingle as a result of their kinetic energy of random motion.

7. A physician orders a dose of vincristine, based on the patient's body weight, as 25 mcg/kg. The maximum dose of vincristine is capped at 2 mg, so if your calculated dose exceeds that, it is automatically reduced to 2 mg. The drug is available as 1 mg/mL vials. The patient weighs 143 pounds. How many mL are used for a dose?

a. 1 mL
b. 1.6 mL
c. 2 mL
d. 3.6 mL

8. Which of the following is a high-alert medication?

   a. sulfamethoxazole and trimethoprim
   b. levofloxacin
   c. clopidogrel
   d. normal saline

9. On a prescription, which part is considered the inscription?

   a. the directions to the patient
   b. the medication ordered
   c. the quantity to dispense
   d. the prescriber's signature

10. What kind of medication formulary is a community pharmacy most likely to have?

   a. closed formulary
   b. open formulary
   c. proprietary formulary
   d. generic formulary

11. If a medication were labeled as expiring on 10/1/2020, on what day does the medication expire?

   a. 1
   b. 15
   c. 30
   d. 31

12. If a prescriber uses a fax machine to send a pharmacy a prescription for a Medicaid patient, are they required to use a tamper-resistant prescription form?

a. Yes, that was a requirement of the Medicaid Tamper-Resistant Prescription Pad Law.

b. No, prescriptions sent via a fax machine are exempt from the Medicaid Tamper-Resistant Prescription Pad Law.

c. The requirements for a fax machine, with respect to the Medicaid Tamper-Resistant Prescription Pad Law, depends as to whether the prescriber is from an inpatient or an outpatient setting.

d. This falls into a gray area in the law that was not considered when the Medicaid Tamper-Resistant Prescription Pad Law was first passed.

13. Which of the following is an example of a public health insurance?

    a. Medicare
    b. Medicaid
    c. SCHIP
    d. all of the above

14. Which anti-infective agent could cause a disulfarim-like reaction (nausea, vomiting, flushing of the skin, tachycardia, and shortness of breath) if they consume alcohol while using it?

    a. nystatin
    b. doxycycline
    c. metronidazole
    d. vancomycin

15. If a medication is listed as expiring 12/2020, on what day does it expire?

    a. November 30, 2020
    b. December 1, 2020
    c. December 31, 2020
    d. January 1, 2021

16. How many milliliters of a 14.6% stock solution should be used to prepare 1 liter of a 0.225% solution?

a. 15.4 mL
b. 64.9 mL
c. 146 mL
d. none of the above

17. Which of the following terms means mishearing or misinterpretation of a phrase as a result of near-homophony?

    a. mondegreen
    b. Manfred Mann
    c. palindrome
    d. homonym

18. Which of the following is an example of risk management?

    a. Carvedilol was accidentally dispensed incorrectly from the pharmacy instead of captopril. The root cause was because the technician accidentally grabbed the wrong unit dose package, as they are both small white tablets that are close to each other in the alphabet. The solution was to create a new policy that each drug will be set physically separate from each other on different shelves.
    b. checking a floor stock list to ensure it has been filled correctly before it leaves the pharmacy
    c. The pharmacy devises a method to make sure that parenteral nutrition solutions above 1000 mOsmol/L are never infused in a peripheral IV line.
    d. having a policy that all chemotherapeutic agents must be checked by at least two pharmacists prior to dispensing

19. Which prescription origin code (POC) should be used if a physician orders a medication through the pharmacy's IVR?

    a. 1
    b. 2
    c. 3
    d. 4

20. How many bottles of erythromycin ethylsuccinate 400 mg tablets should be ordered based on the following information?
    Package size: 100 tablets/bottle
    Minimum number of units: 60 tablets
    Maximum number of units: 240 tablets
    Current inventory level: 57 tablets

    a. 0 bottles
    b. 1 bottle
    c. 2 bottles
    d. 3 bottles

21. If a Medicare patient purchases insulin syringes, which part of their insurance should the syringes be processed under?

    a. Medicare Part A
    b. Medicare Part B
    c. Medicare Part D
    d. The patient will need to pay the cash price.

22. Which of the following are common challenges to creating and maintaining databases?

    a. getting various databases to share information
    b. security
    c. both a and b
    d. There are usually no challenges to creating and maintaining databases.

23. Valtrex (valacyclovir) would be most accurately classified as which of the following?

    a. antibiotic
    b. amebicide
    c. antiviral
    d. viricide

24. Which of the following medications is a schedule IV controlled substance?

a. Xanax
b. Vicodin
c. Tylenol No. 3
d. Flexeril

25. The directions for a bulk vial containing 10 gram of lyophilized vancomycin recommends reconstitution with 95 milliliters of diluent. If the vial has 5 milliliters of powder volume, what will be the resulting concentration in mg/mL after adding the suggested quantity of diluent?

a. 0.01 mg/mL
b. 1 g/10 mL
c. 100 mg/mL
d. 105 mg/mL

26. Which NCC MERP classification does the following situation belong in? A patient received a dose of clarithromycin, when they should have received azithromycin instead. The patient suffered no ill effects from the medication misadventure.

a. no error
b. error, no harm
c. error, harm
d. error, death

27. Which of the following is an example of quality improvement?

a. Carvedilol was accidentally dispensed incorrectly from the pharmacy instead of captopril. The root cause was because the technician accidentally grabbed the wrong unit dose package, as they are both small white tablets that are close to each other in the alphabet. The solution was to create a new policy that each drug will be set physically separate from each other on different shelves.
b. checking a floor stock list to ensure it has been filled correctly before it leaves the pharmacy
c. The pharmacy devises a method to make sure that

parenteral nutrition solutions above 1000 mOsmol/L are never infused in a peripheral IV line.

    d. having a policy that all chemotherapeutic agents must be checked by at least two pharmacists prior to dispensing

28. Which of the following is an example of a signatura?

    a. Lipitor 20 mg
    b. Disp: #30
    c. i tab PO daily
    d. the physician's name signed above a line stating, "Product Selection Permitted"

29. How many bottles of metformin 500 mg tablets should be ordered based on the following information?
Package size: 500 tablets/bottle
Minimum number of units: 200 tablets
Maximum number of units: 800 tablets
Current inventory level: 250 tablets

    a. 0 bottles
    b. 1 bottle
    c. 2 bottles
    d. 3 bottles

30. Which of the following insurance plans is an example of managed care?

    a. HMO
    b. PPO
    c. EPO
    d. all of the above

31. Which of the following is an example of a pharmacy technology that can be utilized to deliver medications from a pharmacy to a nursing unit in an institutional setting?

    a. robotic delivery system
    b. pneumatic tube system
    c. automated dispensing cabinets (also called medstations)

d. all of the above

32. Why do patients often receive tapered doses of glucocorticosteroids?

    a. to ween patients off of their addictive properties
    b. so their bodies slowly start creating their own endogenous source of steroids again
    c. to help mollify the erratic mood swings associated with steroid use
    d. none of the above

33. You receive a prescription for a controlled substance written by a Charles Alderton Pepper, M.D., and it has the following DEA number recorded on it - DP4800620. Does Dr. Pepper's DEA number validate, and if not then why does it not validate?

    a. This DEA number validates.
    b. The letter(s) at the beginning of this DEA number are incorrect.
    c. incorrect check sum
    d. incorrect number of digits

34. A 154 pound patient with a systemic fungal infection is to receive amphotericin B lipid complex 5 mg/kg daily. How many milligrams of amphotericin B should this patient receive daily?

    a. 770 mg
    b. 350 mg
    c. 1694 mg
    d. none of the above

35. Why might Celebrex and Cerebyx be mistaken for each other?

    a. They have near-homophony names (sound-alike).
    b. At quick glance, the names look very similar (look-alike).
    c. Both 'a' and 'b' are possible reasons that the two

medications could be mistaken for each other.
d. No one would ever mistake them for each other.

36. When donning PPE to prepare sterile products, what is the last item you should don?

    a. a nonshedding gown with sleeves that fit snugly around the wrists
    b. sterile gloves
    c. shoe covers
    d. face mask

37. A prescription is written for prochlorperazine 25 mg, Disp #12, Sig: i supp pr q12h prn n/v, Refills NR. What is the intended route of administration for this medication?

    a. orally
    b. rectally
    c. vaginally
    d. intravenously

38. How many bottles of Zofran 4 mg/5 mL oral solution should be ordered based on the following information?
    Package size: 50 milliliters/bottle
    Minimum number of units: 100 milliliters
    Maximum number of units: 400 milliliters
    Current inventory level: 110 milliliters

    a. 0 bottles
    b. 1 bottle
    c. 2 bottles
    d. 3 bottles

39. What is another term for explanation of benefits (EOB)?

    a. bill
    b. remittance advice
    c. payment received
    d. medication therapy management

40. As the quantity of materials that a pharmacy is expected to have on hand to dispense increases over time, which of the following technologies should save the pharmacy time and space?

    a. automated storage and retrieval systems
    b. interactive voice response
    c. automated compounding devices
    d. none of the above

41. If a patient were prescribed both Ditropan XL and Detrol LA, what would it be considered an example of?

    a. synergy
    b. therapeutic duplication
    c. drug-disease contraindication
    d. acetylcholinesterase replacement therapy

42. You receive a prescription for a controlled substance written by a M. C. Feelgood, M.D., and it has the following DEA number recorded on it - AF6543213. Does Dr. Feelgood's DEA number validate, and if not then why does it not validate?

    a. This DEA number validates.
    b. The letter(s) at the beginning of this DEA number are incorrect.
    c. incorrect check sum
    d. incorrect number of digits

43. A patient with Hodgkin's disease is to receive a maintenance dose of bleomycin 10 units/m^2. If the patient is 170 cm tall and weighs 73 kg, how many units of bleomycin should the patient receive?

    a. 1.99 m^2
    b. 19.9 units
    c. 1.86 m^2
    d. 18.6 units

44. Which of the following is not considered an error prevention strategy?

    a. MedWatch
    b. five rights
    c. tallman lettering
    d. ProDURs

45. During the in-take process, which of the following is not considered critical?

    a. who they're rooting for in the Super Bowl
    b. the concurrent use of any OTC and/or dietary supplements
    c. allergies
    d. various conditions the patient is being treated for

46. A prescription is written for cephalexin 250 mg, Disp 28, Sig: i cap po q6h for 7 days, Refills: NR. How many doses of this medication should the patient receive daily?

    a. 1
    b. 4
    c. 6
    d. 7

47. How many bottles of sertraline 50 mg tablets should be ordered based on the following information?
Package size: 100 tablets/bottle
Minimum number of units: 200 tablets
Maximum number of units: 800 tablets
Current inventory level: 120 tablets

    a. 0 bottles
    b. 1 bottle
    c. 6 bottles
    d. 7 bottles

48. A common reject code if a prescriber's NPI has not been

entered into the pharmacy management software system is '25'. What does this reject code mean?

a. product/service not covered
b. refill to soon
c. missing/invalid prescriber ID
d. switch vendor is currently down

49. Some chain pharmacies have started offering innovative new services to improve communication with customers. One such example is having pharmacists that can enter private web-based chat sessions (with or without video) with patients to address concerns and questions over their medication therapy. What kind of technology is that an example of?

a. IVR
b. telepharmacy
c. EHR
d. the annoying kind

50. Which of the following medications is most likely to cause a photosensitivity reaction?

a. Avelox
b. Augmentin
c. Zithromax
d. Vibramycin

51. If controlled substances are stolen from a pharmacy, what form does the DEA require the pharmacy to fill out?

a. The police report is adequate.
b. DEA form 222
c. DEA form 106
d. In the case of theft, the DEA leaves everything to the local authorities and does not gather any data.

52. Which of the following is an example of a PEC that can be used for preparing hazardous drugs?

a. BSC
b. CACI
c. all of the above
d. none of the above

53. Labeling the shelves where tramadol and trazodone are kept as traMADol and traZODone is an example of which kind of strategy to reduce medication errors?

   a. tallman lettering
   b. risk evaluation and mitigation strategy
   c. barcoding technology
   d. separating inventory

54. Pharmacy staff members often have communications with which group(s)?

   a. amongst the various pharmacy staff members
   b. interprofessionally with physicians, nurses, third-party payors, and other professional groups and individuals
   c. communications between the pharmacy staff and the patients
   d. all of the above

55. A prescription is written for amoxicillin susp. 400 mg/5 mL, Disp 100 cc, Sig: i tsp po q12h till all of medicine is gone. How many days' supply does this prescription allow for?

   a. 3
   b. 6
   c. 10
   d. 20

56. Which of the following is/are advantages to wholesaler purchasing when compared to direct purchasing?

   a. reduced turn around time for orders
   b. lower inventory levels and associated costs
   c. reduced commitment of time and staff compared to direct

purchasing
   d. all of the above

57. What type of Medicare health plan is offered by private
    companies that contract with Medicare to provide patients
    with all their Part A and Part B benefits.

   a. Medicare Part A
   b. Medicare Part B
   c. Medicare Part C
   d. Medicare Part D

58. One of HIPAA's goals is to help ensure that heath care
    providers secure a patient's PHI. Which kind of security
    standards are defined by HIPAA?

   a. administrative security of PHI
   b. physical security of PHI
   c. technical safeguards to protect PHI
   d. all of the above

59. Which of the following drug classifications does not carry a
    black box warning about increased risk of suicidal thinking
    and behavior in children, adolescents, and young adults?

   a. tricyclic antidepressants
   b. Selective Serotonin Reuptake Inhibitors (SSRIs)
   c. Serotonin-norepinephrine reuptake inhibitors (SNRIs)
   d. opioid analgesics

60. Why would the FDA require a medication to establish a Risk
    Evaluation and Mitigation Strategy?

   a. This is a common requirement for investigational drugs.
   b. This is a common requirement for newly approved
   medications.
   c. This is due to safety concerns related to using the
   medication.
   d. all of the above

61. Which chapter in the USP provides enforceable regulations for nonsterile compounding?

    a. USP 795
    b. USP 797
    c. USP 1075
    d. USP 1160

62. An adverse event to which of the following should not be reported to MedWatch?

    a. cranberry pills
    b. tadalafil
    c. thiamine
    d. Gardasil

63. If a medication is listed as expiring on 10/2020, what day of the month does it expire on?

    a. 1
    b. 15
    c. 30
    d. 31

64. A prescription is written for Cortisporin Otic 10 mL, Sig: iv gtt ad qid. How many days' supply does this prescription allow for?

    a. 6
    b. 12
    c. 25
    d. 30

65. What may pharmacies request on general purchase orders?

    a. noncontrolled medications
    b. schedule II medications
    c. schedule III-V medications
    d. both a and c

66. Who determines Worker's compensation benefits?

    a. federal government
    b. state government
    c. the judicial branch
    d. no one

67. Which of the following items is most likely to cause a pharmacy to not implement new pharmacy automation or technology?

    a. costs directly and indirectly associated with purchasing and implementing automation and technology
    b. reduction in existing HR salaries and benefits
    c. improved productivity and workflow efficiency
    d. improved operations services

68. For which antipsychotic medication must patients be enrolled in a national registry and closely monitored?

    a. Clozaril
    b. Haldol
    c. Zyprexa Zydis
    d. Geodon

69. Which of the following would not be considered PHI?

    a. 21 cases of measles reported in Newark, Texas in 2013
    b. SSN 333-29-4189, colon cancer
    c. 1815 Metropolitan St, 3 y.o. female with chickenpox
    d. John Powers, BPH

70. Provided a buffer room contains PECs and is not its own PEC, what should the ISO class of the buffer room be?

    a. ISO Class 5 or better
    b. ISO Class 7 or better
    c. ISO Class 8 or better
    d. Class 100 or better

71. Which reporting system is co-managed by the Centers for Disease Control and Prevention (CDC) and the Food and Drug Administration (FDA)?

    a. MedWatch
    b. MERP
    c. VAERS
    d. none of the above

72. A prescription is written for Fosamax 70 mg, Disp: #4, Sig: i tab po q week. How many days' supply does this prescription allow for?

    a. 4
    b. 7
    c. 28
    d. 30

73. How many bottles of zidovudine 300 mg tablets should be ordered based on the following information?
Package size: 60 tablets/bottle
Minimum number of units: 20 tablets
Maximum number of units: 80 tablets
Current inventory level: 24 tablets

    a. 0 bottles
    b. 1 bottle
    c. 2 bottles
    d. 3 bottles

74. What is the NCPDP?

    a. The NCPDP is a nonprofit organization that creates national standards for electronic health care transactions used in prescribing, dispensing, monitoring, managing, and paying for medications and pharmacy services.
    b. The NCPDP is a unique product identifier used for medications intended for human use. It is a unique 10-digit, 3-segment numeric identifier assigned to each medication.
    c. The NCPDP is an identification number assigned to a

particular quantity (or lot) of material from a single manufacturer made in a specific batch.

    d. The NCPDP is a nonprofit organization that accredits more than 20,000 healthcare organizations and programs in the United States.

75. Which of the following technologies can be used to provide health care professionals and/or patients with clinical knowledge and patient-related information?

    a. CDS
    b. IVR
    c. ACD
    d. ADC

76. Lovenox belongs to which pharmaceutical classification?

    a. anticonvulsant
    b. anticoagulant
    c. antibiotic
    d. antipsychotic

77. Which class of drug recall would a product belong to if the violative product may cause temporary or medically reversible adverse health consequences?

    a. class I recall
    b. class II recall
    c. class A recall
    d. class B recall

78. Which of the following is not an example of a medication error?

    a. giving a patient a medication in the morning that was scheduled for the evening
    b. giving a medication intended for PR use PV instead
    c. giving a patient a metoprolol tartrate 50 mg tablet instead of a metoprolol succinate 50 mg capsule
    d. All of the above examples should be considered

medication errors.

79. Some medications need to be kept refrigerated. What temperature range is considered refrigerated according to the USP?

    a. -25° to -10° C
    b. 2° to 8° C
    c. 15° to 30° C
    d. 59° to 86° C

80. When dealing with resistance to change in order to implement new pharmacy automation and technology, who will typically present this resistance?

    a. pharmacy staff
    b. prescribers and other health care team members that will be impacted
    c. patients
    d. all of the above

81. Which of the following drugs is an antiarrhythmic?

    a. Micardis
    b. Multaq
    c. Monopril
    d. Minocin

82. How often should documented cleanings of the counters in a buffer room occur, provided that the counters do not function as the primary engineering control?

    a. per shift
    b. daily
    c. weekly
    d. monthly

83. The abbreviation "U" should be avoided because it can be misinterpreted as which of the following when poorly written?

a. 0
b. 4
c. cc
d. all of the above

84. A prescription is written for metoprolol tartrate 25 mg tabs, Disp: #100, Sig: ss tab po bid, Refills: 2. How should this prescription be adjusted to provide the medication in 30 day increments?

   a. dispense 30 tabs per fill with 9 refills and no partial fills
   b. dispense 30 tabs per fill with 10 refills and no partial fills
   c. dispense 60 tabs per fill with 4 refills and no partial fills
   d. dispense 30 tabs per fill with 5 refills and no partial fills

85. During which cell phase cycle do antineoplastic agents not work?

   a. S
   b. G_2
   c. G_1
   d. G_0

86. Which organization establishes the requirements for drug information resources at various pharmacies?

   a. FDA
   b. OSHA
   c. each state's board of pharmacy
   d. MSDS

87. Placing sulfadiazine and sulfasalazine on different shelving units is an example of which kind of strategy to reduce medication errors?

   a. tallman lettering
   b. risk evaluation and mitigation strategy
   c. barcoding technology
   d. separating inventory

88. If a pharmacy maintains hard copies of their prescriptions, which prescriptions should be marked with a red colored 'C'?

    a. CIII medications
    b. CIV medications
    c. CV medications
    d. all of the above

89. Which of the birth control medications is available as an intrauterine device?

    a. Mirena
    b. Apri
    c. Yaz
    d. DepoProvera

90. A prescription is written for Spiriva, Disp: 1 box c 30 capsules, Sig: inhale contents of i cap qd using HandiHaler. How many days' supply does this prescription allow for?

    a. 30
    b. 60
    c. 90
    d. none of the above

Answers to practice exam 2

1. C	12. B	23. C
2. B	13. D	24. A
3. B	14. C	25. C
4. C	15. C	26. B
5. D	16. A	27. C
6. A	17. A	28. C
7. B	18. A	29. A
8. C	19. B	30. D
9. B	20. B	31. D
10. B	21. B	32. B
11. A	22. C	33. B

34.B	53.A	72.C
35.C	54.D	73.A
36.B	55.C	74.A
37.B	56.D	75.A
38.A	57.C	76.B
39.B	58.D	77.B
40.A	59.D	78.D
41.B	60.C	79.B
42.C	61.A	80.D
43.D	62.D	81.B
44.A	63.D	82.B
45.A	64.B	83.D
46.B	65.D	84.A
47.C	66.B	85.D
48.C	67.A	86.C
49.B	68.A	87.D
50.D	69.A	88.D
51.C	70.B	89.A
52.C	71.C	90.A

CHAPTER 11 After You're Certified

Once you've taken the Pharmacy Technician Certification Exam and successfully passed it, you may wonder, "What's next?". This chapter will attempt to answer that question.

Attaboy/attagirl

All of your hard work has paid off and you deserve a congratulations. Whether it involves breaking into your own personal happy dance, calling a friend or family member, going out for a nice meal, or enjoying a quiet evening where you don't need to study, a celebration of your efforts has been well earned.

Update that résumé

You now officially have the title CPhT, the abbreviation for certified pharmacy technician. This title should be placed with your name at the top of your resume, and should be noted in a section you have listed for certifications. Regardless of the requirements for certification in your state, potential employers will want to know if you are certified, as it demonstrates a commitment to your field, and it assures them of a level of knowledge that you have achieved. By providing those initials, CPhT, beside your name at the top of your résumé, it immediately signals them that you are certified. Also, by including your certification number in an easily identifiable subsection of your résumé, it makes it easier for a potential employer to verify that you are certified.

Does your state want to know that you are certified?

As pharmacy technicians are currently (October 2013) regulated on the state level, requirements for working as a pharmacy technician vary widely from state to state. This section is intended to give you a brief overview of what the various states require for someone to work as a pharmacy technician. Please keep in mind that this section is not intended as a replacement for checking with an individual state board of pharmacy to verify their requirements. A good resource for links to each board of pharmacy is the National Association of Boards of Pharmacy (www.nabp.net/boards-of-pharmacy).

It is also noteworthy that any costs listed for state registrations and/or renewals do not include any additional costs associated - background checks (if required) or late fees and penalties.

Alabama

The state of Alabama has the following requirements for pharmacy technicians:

- Pharmacy technicians must be at least 17 years of age.
- Pharmacy technicians are required to register ($60) with the state, and they must renew ($60) their registration biennially by October 31st of odd numbered years (2015, 2017, 2019, etc.).
- Pharmacy technicians are required to complete three hours of continuing education annually, one of which must be a live presentation.
- The state of Alabama does not require certification, but a pharmacy is allowed to increase their technician to pharmacist ratio from 2:1 up to 3:1 if one of the technicians is certified.

Alaska

The state of Alaska has the following requirements for pharmacy technicians:

- Pharmacy technicians must be at least 18 years of age, and have a high school diploma or equivalent.
- Pharmacy technicians are required to register ($150) with the state, and they must renew ($100) their registration biennially by June 30th of even numbered years (2014, 2016, 2018, etc.).
- Pharmacy technicians are required to complete ten hours of continuing education per renewal period.
- Certification is not required in the state of Alaska.

Arizona

The state of Arizona has the following requirements for pharmacy technicians:

- Pharmacy technicians must be at least 18 years of age and have a high school diploma or equivalent.
- Pharmacy technicians are required to register ($46--prorated) with the state, and they must renew ($36) their registration annually by November 1st.
- Pharmacy technicians are required to complete twenty hours of continuing education biennially, two of which must be in law.
- Completion of a board approved training program and board approved certification are required in the state of Arizona.

Arkansas

The state of Arkansas has the following requirements for pharmacy technicians:

- Pharmacy technicians must have a high school diploma or equivalent.
- Pharmacy technicians are required to register ($108.50) with the state, and they must renew ($35) their registration

annually.
- Certification is not required in the state of Arkansas.

California

The state of California has the following requirements for pharmacy technicians:

- Pharmacy technicians must have a high school diploma or equivalent.
- Pharmacy technicians are required to register ($80) with the state, and they must renew ($100) their registration annually.
- A pharmacy technician is required to have either completed an appropriately accredited training program, or passed a board approved certification.

Colorado

The state of Colorado does not require certification, but a pharmacy is allowed to increase their technician to pharmacist ratio from 2:1 up to 3:1 if one of the technicians has accomplished any one of the following:

- is nationally certified,
- has a degree from an accredited pharmacy technician program,
- or has completed 500 hours of experiential training.

Connecticut

The state of Connecticut has the following requirements for pharmacy technicians:

- Pharmacy technicians are required to register ($100) with the state, and they must renew ($50) their registration annually.
- Pharmacy technicians are required to complete twenty hours of continuing education biennially, two of which must be in law.

- The state of Connecticut does not require certification, but a pharmacy is allowed to increase their technician to pharmacist ratio from 2:1 up to 3:1 if the technician is either certified by a board approved program, or if the pharmacy prepares sterile products, dispenses unit doses, and does bulk compounding.

Delaware

The state of Delaware does not have any requirements for pharmacy technicians.

District of Columbia

The District of Columbia (Washington D.C.) does not have any requirements for pharmacy technicians.

Florida

The state of Florida has the following requirements for pharmacy technicians:

- Pharmacy technicians are required to register ($105) with the state and they must renew ($55) their registration biennially by December 31st of even numbered years (2014, 2016, 2018, etc.).
- Pharmacy technicians are required to complete a minimum of 20 hours of continuing education, of which 4 hours must be via live presentation, and 2 hours must be related to the prevention of medication errors. During a pharmacy technicians first renewal, 1 hour must be in HIV/Aids education.
- Completion of a board approved training program is required in the state of Florida.

Georgia

The state of Georgia has the following requirements for pharmacy technicians:

- Pharmacy technicians must be at least 17 years of age, and is either pursuing a high school diploma, or has a high school diploma or equivalent.
- Pharmacy technicians are required to register ($100) with the state, and they must renew ($60) their registration biennially by June 30th of odd numbered years (2015, 2017, 2019, etc.).
- Certification is not required, but a pharmacy is allowed to increase their technician to pharmacist ratio from 2:1 up to 3:1 if one of the technicians is nationally certified.

Guam

The territory of Guam does not have any requirements for pharmacy technicians.

Hawaii

The state of Hawaii does not have any requirements for pharmacy technicians.

Idaho

The state of Idaho has the following requirements for pharmacy technicians:

- Pharmacy technicians must be at least 18 years of age, and have a high school diploma or equivalent.
- Pharmacy technicians are required to register ($35) with the state, and they must renew ($35) their registration annually by June 30th.
- Pharmacy technicians in the state of Idaho must be nationally certified.

Illinois

The state of Illinois has the following requirements for pharmacy technicians:

- Pharmacy technicians must be at least 16 years of age, and is either pursuing a high school diploma, or has a high school diploma or equivalent.
- Pharmacy technicians are required to register ($40) with the state, and they must renew ($25) their registration annually by March 31st.
- An individual is not allowed to work as a pharmacy technician for more than two years without obtaining training from a nationally accredited school and passing a national certification exam.

Indiana

The state of Indiana has the following requirements for pharmacy technicians:

- Pharmacy technicians must be at least 18 years of age, and have a high school diploma or equivalent.
- Pharmacy technicians are required to register ($35) with the state, and they must renew ($35) their registration biennially by June 30th of even numbered years (2014, 2016, 2018, etc.).
- Pharmacy technicians in the state of Indiana must either complete a board approved program or complete a board recognized certification within one year of initial registration.

Iowa

The state of Iowa has the following requirements for pharmacy technicians:

- Pharmacy technicians must be at least 18 years of age, and have a high school diploma or equivalent.
- Pharmacy technicians are required to register ($30) with the state, and they must renew ($44) their registration biennially

before the last day of the registrant's birthday.
- Pharmacy technicians in the state of Iowa must complete national certification within one year of initial registration.

Kansas

The state of Kansas has the following requirements for pharmacy technicians:

- Pharmacy technicians are required to register ($50) with the state, and they must renew ($25) their registration annually by October 31st.
- Certification is not required, but a pharmacy is allowed to increase their technician to pharmacist ratio from 2:1 up to 3:1 if two of the technicians have passed a state approved certification exam.

Kentucky

The state of Kentucky has the following requirements for pharmacy technicians:

- Pharmacy technicians must be at least 16 years of age.
- Pharmacy technicians are required to register ($25) with the state, and they must renew ($25) their registration annually by March 31st.

Louisiana

The state of Louisiana has the following requirements for pharmacy technicians:

- Pharmacy technicians must be at least 18 years of age, and have a high school diploma or equivalent.
- An individual that wants to become a pharmacy technician must complete a Pharmacy Technician Candidate Registration application ($25) and then complete 600 hours of training under a licensed pharmacist; after which, they must take the Pharmacy Technician Certification Exam.

- Once an individual has completed their training and passed the Pharmacy Technician Certification Exam, they are required to register ($100) with the state, and they must renew ($50) their registration annually by June 30th.
- Pharmacy technicians are required to complete ten hours of continuing education annually.
- The technician to pharmacist ratio is 2:1 if there is a technician candidate on duty; but the ratio may be increased to 3:1 if there are no technician candidates on duty.

Maine

The state of Maine has the following requirements for pharmacy technicians:

- Pharmacy technicians are required to register ($25) with the state, and they must renew ($25) their registration annually by December 31st.
- Pharmacy technicians that pass a national certification exam are allowed to register with the state as advanced pharmacy technicians.
- The technician to pharmacist ratio is 3:1; however, the pharmacy may increase the ratio to 4:1 if at least one is registered as an advanced pharmacy technician.

Maryland

The state of Maryland has the following requirements for pharmacy technicians:

- Pharmacy technicians must be pursuing a high school diploma, or have a high school diploma or equivalent.
- Pharmacy technicians are required to register ($45) with the state and they must renew ($45) their registration biennially.
- Pharmacy technicians must either complete a 160 hour training program or be nationally certified.
- Pharmacy technicians are required to complete ten hours of continuing education during their first renewal cycle, and twenty hours each subsequent renewal period.

Massachusetts

The state of Massachusetts has the following requirements for pharmacy technicians:

- Pharmacy technicians must be at least 18 years of age, and either pursuing a high school diploma, or have a high school diploma or equivalent.
- Pharmacy technicians are required to register ($136) with the state, and they must renew ($51) their registration biennially by the pharmacy technician's birthday.
- Pharmacy technicians must either complete a board approved training program, or pass a state board approved certification.

Michigan

The state of Michigan does not have any requirements for pharmacy technicians.

Minnesota

The state of Minnesota has the following requirements for pharmacy technicians:

- Pharmacy technicians must be at least 18 years of age, and have a high school diploma or equivalent.
- Pharmacy technicians are required to register ($35) with the state, and they must renew ($20) their registration biennially by January 1st.
- Pharmacy technicians are required to either complete a state approved training program.
- Pharmacy technicians are required to complete twenty hours of continuing education during each renewal cycle.

Mississippi

The state of Mississippi has the following requirements for pharmacy technicians:

- Pharmacy technicians must be at least 18 years of age, and have a high school diploma or equivalent.
- Pharmacy technicians are required to register ($50) with the state, and they must renew ($50) their registration annually between April 1st and March 31st.
- Pharmacy technicians are required to pass a national certification exam prior to their first registration renewal.

Missouri

The state of Missouri requires pharmacy technicians to register ($35) with the state. At the time this book went to print there were no requirements for renewal.

Montana

The state of Montana has the following requirements for pharmacy technicians:

- Pharmacy technicians must be at least 18 years of age, and have a high school diploma or equivalent.
- Pharmacy technicians initially register ($60) as a technician-in-training, and have 18 months in which they must pass a national certification exam. Once certified, registration renewal ($50) is required annually by June 30th.

Nebraska

The state of Nebraska has the following requirements for pharmacy technicians:

- Pharmacy technicians must be at least 18 years of age, and have a high school diploma or equivalent.
- Pharmacy technicians are required to register ($25 + $1 for each year till next renewal cycle) with the state, and they must renew ($25 + $2) their registration biennially by January 1st of odd numbered years (2015, 2017, 2019, etc.).

Nevada

The state of Nevada has the following requirements for pharmacy technicians:

- Pharmacy technicians must be at least 18 years of age, and have a high school diploma or equivalent.
- Pharmacy technicians are required to register ($40) with the state, and they must renew ($40) their registration biennially by October 31st of even numbered years (2014, 2016, 2018, etc.).
- Pharmacy technicians are required to have completed a board approved pharmacy technician training program (including ASHP accredited programs from other states), or complete 1,500 hours as a technician-in-training, or be already licensed in another state, or have completed a training program in another state which is not ASHP accredited, but has also successfully passed the Pharmacy Technician Certification Exam. (As ASHP is transferring its accreditation program to PTAC in 2014, it is expected that the language in the law will be shifted to reflect that.)
- Technicians are required to complete 12 hours of in-service, and 1 hour of continuing education in Nevada law prior to registration renewal.

New Hampshire

The state of New Hampshire has the following requirements for pharmacy technicians:

- Pharmacy technicians must be at least 18 years of age, and must be pursuing a high school diploma or equivalent, or have a high school diploma or equivalent.
- Pharmacy technicians are required to register ($25) with the state, and they must renew ($25) their registration annually by March 31st.

New Jersey

The state of New Jersey has the following requirements for pharmacy technicians:

- Pharmacy technicians must be at least 18 years of age, and have a high school diploma or equivalent.
- Pharmacy technicians are required to purchase an application for registration ($50), and then register ($35 for each year till next renewal cycle) with the state, and they must renew ($70) their registration biennially.

New Mexico

The state of New Mexico has the following requirements for pharmacy technicians:

- Pharmacy technicians initially register ($30) as a non-certified technician, and have 12 months in which they must pass a national certification exam.
- Once certified, the certified pharmacy technician must register ($30) with the state, and renewal ($30) is required biennially by the last day of the registrant's birth month.

New York

The state of New York does not have any requirements for pharmacy technicians.

North Carolina

The state of North Carolina has the following requirements for pharmacy technicians:

- Pharmacy technicians must have a high school diploma or equivalent.
- Pharmacy technicians are required to either complete a board approved 6 month training program within the pharmacy, or a pharmacy technician training program from a community college.

- Pharmacy technicians must register ($30) with the state, and renewal ($30) is required annually by December 31st.
- The state of North Carolina does not require certification, but a pharmacy is allowed to increase their technician to pharmacist ratio from 2:1, as long as each additional technician is nationally certified by the Pharmacy Technician Certification Board.

North Dakota

The state of North Dakota has the following requirements for pharmacy technicians:

- Pharmacy technicians must complete an ASHP acrredited training program. If this training program is on-the-job, they will need to register as a technician-in-training. (As ASHP is transferring its accreditation program to PTAC in 2014, it is expected that the language in the law will be shifted to reflect that.)
- Pharmacy technicians must pass and maintain the Pharmacy Technician Certification Exam.
- Pharmacy technicians must register ($35) with the state, and renewal ($35) is required annually by March 1st.
- Pharmacy technicians are required to complete ten hours of continuing education per renewal period.

Ohio

The state of Ohio has the following requirements for pharmacy technicians:

- Pharmacy technicians must be at least 18 years of age and have a high school diploma or equivalent.
- Pharmacy technicians must pass a board approved certification exam.

Oklahoma

The state of Oklahoma has the following requirements for pharmacy technicians:

- Pharmacy technicians must first complete a training program (Phase I), and then they must complete on-the-job training (Phase II) within 90 days.
- Pharmacy technicians are required to register ($40) with the state, and they must renew ($40) their registration annually by the last day of the registrant's birth month.
- While national certification is not mandatory, only certified technicians may prepare chemotherapy or prepare sterile products with multiple drugs.

Oregon

The state of Oregon has the following requirements for pharmacy technicians:

- Pharmacy technicians must have a high school diploma or equivalent.
- Pharmacy technicians initially register ($50) with the state, and they have 12 months in which they must pass a national certification exam.
- Once certified, pharmacy technicians must renew ($50) their registration annually by August 31st.
- Technicians are required to complete 1 hour of continuing education in Oregon law prior to registration renewal.

Pennsylvania

The state of Pennsylvania does not have any requirements for pharmacy technicians.

Puerto Rico

The territory of Puerto Rico does not have any requirements for pharmacy technicians.

Rhode Island

The state of Rhode Island has the following requirements for pharmacy technicians:

- Rhode Island recognizes two levels of pharmacy technicians. Pharmacy Technician I is considered an employer-specific license; whereas, a Pharmacy Technician II license allows the technician to be automatically eligible for employment at all pharmacies.
- Pharmacy technicians must be at least 18 years of age, and in order to be a Pharmacy Technician II, they must have a high school diploma or equivalent.
- Pharmacy technicians are required to register ($40) with the state, and they must renew ($40) their registration annually on the last day of the registrant's birth month.
- Once a pharmacy technician passes a national certification exam, they are eligible to be a Pharmacy Technician II.

South Carolina

The state of South Carolina has the following requirements for pharmacy technicians:

- Pharmacy technicians must have a high school diploma or equivalent.
- Pharmacy technicians are required to register ($40) with the state, and they must renew ($15) their registration annually by June 30th.
- Pharmacy technicians are required to complete ten hours of continuing education annually, four of which must be a live presentation.
- The state of South Carolina allows a registered certified pharmacy technician to perform more duties than a registered pharmacy technician including: accept verbal orders, perform transfers, check another technician's refill, and check another technician's medication repackaging from a bulk container to unit dose containers.
- To become certified in South Carolina, a technician must: complete an ASHP accredited program, pass the Pharmacy

Technician Certification Exam, and complete 1000 hours of practice under the supervision of a licensed pharmacist. (As ASHP is transferring its accreditation program to PTAC in 2014, it is expected that the language in the law will be shifted to reflect that.)

South Dakota

The state of South Dakota requires pharmacy technicians to register ($25) with the state within their first thirty days of employment, and they must renew ($25) their registration annually by October 31st.

Tennessee

The state of Tennessee requires pharmacy technicians to register ($50) with the state prior to employment, and they must renew ($50) their registration biennially from the date the registration was granted.

Texas

The state of Texas has the following requirements for pharmacy technicians:

- Pharmacy technicians must have a high school diploma or equivalent.
- Pharmacy technicians may initiate employment as a trainee registered with the state, but they may only maintain that status for two years. The application fee for this is $47.
- Pharmacy trainees must pass a state approved exam (currently, the Pharmacy Technician Certification Exam is the only state approved exam) to register as a pharmacy technician.
- Pharmacy technicians not designated as trainees are required to register ($75) with the state, and they must renew ($71) their registration biennially.
- Pharmacy technicians must complete twenty hours of continuing education during each renewal cycle.

Utah

The state of Utah has the following requirements for pharmacy technicians:

- Pharmacy technicians are required to register ($100) with the state, and they must renew ($47) their registration biennially by September 30th of odd numbered years (2015, 2017, 2019, etc.).
- Pharmacy technicians must complete either an on-the-job training program or a formal training program.
- Pharmacy technicians are required to pass and maintain national certification.

Vermont

The state of Vermont requires pharmacy technicians to register ($50) with the state, and they must renew ($50) their registration biennially by July 31st of odd numbered years (2015, 2017, 2019, etc.).

Virgin Islands

The Virgin Islands do not have any requirements for pharmacy technicians.

Virginia

The state of Virginia has the following requirements for pharmacy technicians:

- Pharmacy technicians must either pass the Pharmacy Technician Certification Exam, or complete a state approved training program and pass a state approved exam.
- Pharmacy technicians are required to register ($25) with the state, and they must renew ($25) their registration annually.
- Pharmacy technicians are required to complete five hours of continuing education each renewal period.

Washington

The state of Washington has the following requirements for pharmacy technicians:

- Pharmacy technicians are required to register ($60) with the state, and they must renew ($50) their registration annually before their birthday.
- Pharmacy technicians must complete a state approved training program.
- Pharmacy technicians must pass and maintain a national certification exam.
- Pharmacy technicians must complete four hours of training on HIV/AIDS prior to being registered with the state.

West Virginia

The state of West Virginia has the following requirements for pharmacy technicians:

- Pharmacy technicians must be at least 18 years of age, and have a high school diploma or equivalent.
- Pharmacy technicians must either complete a state approved training program (which could include a 2,080 hour pharmacy trainee program) or become nationally certified and complete 20 hours of training.
- Pharmacy technicians are required to register ($25) with the state, and they must renew ($30) their registration annually by July 1st.

Wisconsin

The state of Wisconsin does not have any requirements for pharmacy technicians.

Wyoming

The state of Wyoming has the following requirements for pharmacy technicians:

- Pharmacy technicians must be at least 18 years of age, and have a high school diploma or equivalent.
- Pharmacy technicians may register as a technician in training ($15) for up to two years and is nonrenewable, and to continue working beyond that, must pass a board approved certification.
- Pharmacy technicians are required to register ($50) with the state, and they must renew ($50) their registration annually by December 31st.
- Pharmacy technicians must complete six hours of continuing education during each renewal cycle.

Set-up CPE Monitor

The Accreditation Council for Pharmacy Education (ACPE) and the National Association of Boards of Pharmacy (NABP) have developed a continuing pharmacy education (CPE) tracking service, CPE Monitor, that will authenticate and store data for completed continuing education received by pharmacists and pharmacy technicians. Most boards of pharmacy require the use of CPE Monitor to track and verify the continuing education credits, and in the future, CPE Monitor will likely be able to flow information to the Pharmacy Technician Certification Board (PTCB), making it easier to renew your certification.

To get started with CPE Monitor, you will first want to proceed to www.mycpemonitor.net (the link will likely redirect you to a web page on NABP's website). The web page will provide some basic information about CPE Monitor. On the web page, you will find a link for 'CPE Monitor Log In'. If you follow that link, it will provide you with an opportunity to create an account. The site does offer answers to frequently asked questions, in case you have any concerns about setting up your profile or about how to use CPE Monitor.

Continuing education

As discussed in the introduction in this book, once certification through the Pharmacy Technician Certification Exam (PTCE) is achieved, certification renewal is required every two years. Pharmacy technicians are required to complete 20 hours of continuing education (CE) every two years. At least one of the 20 hours of CE must be in the area of pharmacy law during each renewal period, and starting in 2014, one of the CE must be in medication safety during each renewal period. In 2015, all CE will need to be pharmacy technician specific (until then, CE designated for pharmacists can still be used by technicians).

Acceptable pharmacy technician continuing education can be obtained through a variety of methods, such as through pharmacy or pharmacy technician professional organizations, as well as through a number of sources online.

You may use 1 college course during your two-year certification period. The college course must be in either a Life Science (Chemistry, Biology, Anatomy, etc.) or Math, and currently counts as 15 hours of CE. That value will be reduced to 10 hours of CE in 2016.

Most popular sources of CE are either web-based or printed. If you need to complete a specific number of live CE, traditionally, you needed to find what was available in your own community or travel to a conference, but it has become increasingly easy to also find online CE course that requires real time interaction with the presenter. The following is not a definitive list of CE resources, but is simply meant to provide a starting point.

- POWER-PAK C.E. offers a broad range of both free and paid for CE, www.powerpak.com
- Pharmacy Times provides a combination of both free and paid for CE, www.pharmacytimes.com
- Free CE offers numerous free articles, videos, and live webinars (this last category can be used for live CE), www.freece.com
- RxSchool is a great resource for online live CE with varying levels of cost, www.rxschool.com
- Continuing Education offers numerous CE with an emphasis on helping professionals meet the requirements set by the states they live in, www.continuingeducation.com/pharmacy
- Tech Lectures offers affordably priced spiral bound CE,

tlectures.com
- The ASHP offers a wide range of both printed and web-based CE, along with webinars and opportunities to attend various conferences to take part in live CE, eleraning.ashp.org/catalog
- The National Pharmacy Technician Association, which also publishes the Today's Technician magazine, offers many CE specifically tailored to pharmacy technicians in both print and web based format. The CE activities are usually free or available at a reduced price for members, www.pharmacytechnician.org
- Pharmacy Technician's Letter is a low cost subscription that provides material both online and in printed format, and the CE are always specific to pharmacy technicians, pharmacytechniciansletter.therapeuticresearch.com
- U. S. Pharmacist offers both free and paid for web-based CE, www.uspharmacist.com

Certification renewal

When you become certified, your certification will be good for two years. Your certificate stating that you are certified will include an expiration date. You can renew your certification (or recertify) up to 100 days prior to when your current certification expires. It is recommended that you recertify at least 30 days prior to your certification's expiration date to provide adequate processing time. The cost to recertify is $40.

To renew your certification you will need to proceed to ptcb.org, and log in using your existing account (which was created when you registered for the exam). Once logged in, you will want to click the link titled 'Apply for Recertification' and follow the direction on the screen.

If you want further instructions, or have trouble logging in, please check out the Recertification Guide that the Pharmacy Technician Certification Board (PTCB) created. You may checkout PTCB's Recertification Guide at http://goo.gl/rMPRGk (for your convenience, Google's URL shortener has been used).

Additional certifications to consider

As the career paths available to pharmacy technicians continue to grow and become more complex, additional certifications may help technicians advance their careers further. The current certifications available for technicians to further their careers beyond passing either the PTCE or ExCPT include sterile products certification, chemotherapy certification, and compounding (nonsterile) certification. A certification in sterile compounding is intended to advance a technician's or pharmacist's knowledge in preparing sterile products while using proper aseptic technique and their knowledge of USP 797 to ensure their ability to comply with all appropriate guidelines. Chemotherapy certification is intended to provide both pharmacists and technicians with the necessary skills and knowledge to safely prepare various hazardous drugs. A general compounding certification for technicians and pharmacists focuses on nonsterile extemporaneous compounding as both an art and a science. All three of these certifications include both didactic and hands-on portions; and as various institutions and organizations offer these certifications, the prices of obtaining these certifications can vary greatly.